Comments by Author

I write because there is no planet B.

Any similarity between the author's parents and the protagonist's parents is purely intentional.

Thank you to my Beta readers

You know who you are

NOT ALONE ON EARTH

Eloise Hamann

Not Alone on Earth
Third in the *Ocean Worlds Trilogy*
Copyright © 2025 by Eloise Hamann

ISBN: 979-8-9898693-0-5

Cover design: David Donovan
Photographer: Douglas McDonald
Interior layout: Val Sherer, Personalized Publishing Services

All rights reserved.
Printed in the United States of America

Contents

Prologue, *1*

Chapter 1
(Elizabeth's surprise discovery), 3

Alvin Chapter 2
(The alien Elizabeth saw), 8

Chapter 3
(Potential problem in the nursery), 11

Alvin Chapter 4
(The second encounter), 19

Chapter 5
(Elizabeth spots the alien again), 20

Alvin Chapter 6
(Alvin relates a secret to Fitz), 27

Chapter 7
(Elizabeth seeks a name for the alien), 31

Chapter 8
(Pierre and Elizabeth connect), 33

Chapter 9
(One month earlier), 37

Chapter 10
(The Naming), 41

Chapter 11
(Bizarre history of Elizabeth's hometown), 48

Chapter 12
(Trouble on Yellow Submarine), 51

Chapter 13
(Elizabeth and Tyrone have an astonishing encounter), 55

Chapter 14
(Death on Yellow Submarine), 58

Chapter 15
(The source of the dreaded disease), 63

Chapter 16
(The fight to contain rats; Alon shares his personal story), 68

Chapter 17
(The tense meeting), 72

Chapter 18
(Pierre and Alessandra meet), 75

Chapter 19
(Three on the caterpillar), 77

Chapter 20
(Sanjay's condition, Team meeting), 81

Chapter 21
(Elizabeth ponders and reminisces), 85

Chapter 22
(Conference in the Situation Room), 87

Chapter 23
(Tyrone describes 'the talk'), 89

Chapter 24
(Alon's history), 92

Chapter 25
(Trouble in the nursery again?), 94

Chapter 26
(Elizabeth Muses About Long-Term Change), 97

Chapter 27
(Alon's unusual request), 100

Chapter 28
(Fight over leadership of ROE), 103

Chapter 29
(Elizabeth tries to share information with Alon), 106

Chapter 31
(Command Center Surprise), 112

Alvin Chapter 32
(Reaction to Jackson's chase), 119

Chapter 33
(Director Jackson threatens trouble), 121

Chapter 34
(Elizabeth begins to plot), 125

Chapter 35
(Elizabeth's concern for Alon grows), 129

Alvin Chapter 36
(Alvin makes a most unpleasant discovery), 131

Chapter 37
(Sanjay and Jackson debate), 134

Chapter 38
(Elizabeth frets), 137

Chapter 39
(Elizabeth plots before her departure), 140

Chapter 40
(Elizabeth sets out on her leave), 143

Chapter 41
(Interlude in Perth before traveling to NYC), 145

Chapter 42
(Arrival at the hotel near the UN building in NYC), 148

Chapter 43
(Debriefing with Pierre and Hans), 154

Chapter 44
(Elizabeth polishes her Alon report), 158

Chapter 45
(Elizabeth and Pierre tour NYC), 161

Chapter 46
(Elizabeth and Pierre at Dinner), 166

Chapter 47
(Elizabeth receives bad news), 169

Chapter 48
(Elizabeth travels to her roots), 172

Chapter 49
(Elizabeth's mother's time is limited), 175

Chapter 50
(Violet Gayer is no longer suffering from Alzheimer's), 182

Chapter 51
(Dry Creek), 187

Chapter 52
(Elizabeth and siblings arrange the services), 190

Chapter 53
(Thank you notes and memories), 193

Chapter 54
(Elizabeth returns to NYC), 197

Chapter 55
(Elizabeth launches her campaign), 200

Chapter 56
(Elizabeth & Pierre explore the wonders of NYC), 203

Chapter 57
(The wonderful day is not yet over), 206

Chapter 58
(Elizabeth's campaign begins), 209

Chapter 59
(Meetings with Congress members), 211

Chapter 60
(Elizabeth gets a break from an unexpected source), 213

Chapter 61
(Elizabeth on the What's Up *Show), 216*

Chapter 62
(Director Jackson summons Elizabeth), 219

Chapter 63
(Elizabeth returns to Yellow Submarine), 221

Chapter 64
(Elizabeth and Tyrone's encounter), 226

Chapter 65
(The caterpillar is in trouble), 230

Chapter 66
(Post-disaster meeting), 238

Chapter 67
(Back to NYC to launch her campaign), 241

Chapter 68
(Elizabeth schemes with Haihong), 247

Chapter 69
(Elizabeth meets with Jackson), 249

Chapter 70
(Elizabeth is on fire), 253

Chapter 71
(Back home on Yellow Submarine), 255

Chapter 72
(Elizabeth learns shocking news), 258

Chapter 73
(Helen Keller's secret remains secret), 263

Chapter 74
(Command Center Meeting), 265

Chapter 75
(Elizabeth waits), 270

Chapter 76
(Things calm down for Elizabeth), 273

Chapter 77
(NSF Board meeting on The Alien Project), 275

Chapter 78
(Jackson doesn't give up), 278

Chapter 79
(Sanjay works on Command Center), 279

Chapter 80
(Decision on Elizabeth's reinstatement and Sanjay's return), 281

Chapter 81
(Elizabeth makes progress with Alon), 283

Alvin Chapter 82
(Alvin reminisces), 287

Chapter 83
(Elizabeth continues to work on communication with Alon), 289

Chapter 84
(Alon is making progress), 293

Chapter 85
(Elizabeth solicits Sanjay's help), 295

Chapter 86
(Elizabeth warns Alon), 297

Alvin Chapter 87
(Alvin's dilemma), 299

Alvin Chapter 88
(Alvin visits the leaders of the Blues), 300

Chapter 89
(Yusuf supports Jackson), 304

Chapter 90
(Congressional House considers funds for Jackson), 307

Chapter 91
(Follow the Money), 311

Chapter 92
(Elizabeth garners information on Alonites), 313

Chapter 93
(The ROE team weighs in on topics relevant to Alonite life), 315

Chapter 94
(Jackson's significant announcement), 319

Alvin Chapter 95
(Alvin and Haggard encounter a problem), 320

Chapter 96
(Jackson visits Yellow Submarine), 325

Chapter 97
(Jackson stops at Yellow Submarine), 328

Chapter 98
(ROE team anticipates Jackson's story), 330

Chapter 99
(Jackson addresses ROE team), 332

Epilogue; 336

Relevant websites, 338

About the Author, 339

Look deep into nature, and then you will understand everything better.
— Albert Einstein

Prologue

The year is 2033. The world failed to keep Earth from warming beyond two degrees centigrade by the end of the 2020s. Severe disasters are occurring more frequently. The United Nations, at the advice of its COP 30 meeting on climate change, explored livable options for the remaining human population. Many scientists had envisioned settling on Mars. However, the expense of making suitable homes and cities for humans and transporting people with supplies made this infeasible for the immediate future. Sensible scientists asked, "Why are we talking about going up when it would be easier to move down? Oceans cover 71% of our Earth. Why not design submarines to serve as cities?" While many formulated reasons why this wouldn't work, no one opposed an experiment with a modest-sized model.

The United Nations, individual corporations, concerned entrepreneurs, and foundations pitched in to fund project Yellow Submarine in honor of the Beatles, a favorite of the elderly Secretary General.

Yellow Submarine would rest in the Indian Ocean west of Perth, the capital of Western Australia. The depth would be below the lowest level that military and other submarines had ever traveled. With Freemantle Harbour, several airports and a historic site from WWII nearby, Perth provided an ideal place. The competition for the siting of the submarine had been strong. The US settled with housing the submarine's oversight committee titled Command Center in the UN Building in New York City.

In addition to doctors, dentists, librarians, other professionals, and everyday workers required for a modern city, eleven scientists, titled the ROE team, were selectively chosen for Research, Oversight, and Exploration purposes.

Chapter 1

(Elizabeth's surprise discovery)

Elizabeth stormed into the scientists' lounge/dining room shortly before a meeting of the ROE team. "All right, who the hell turned off the recorder on the front monitor when they took out the last recording?" Elizabeth had called their bathyscaphe the caterpillar because the shape reminded her of the caterpillars that lived in the milkweeds near her family's farm.

Everyone looked up, startled by Elizabeth's outburst. "Oh, so someone bumped it off when they removed the recording, and we missed a video of a skate and maybe even an anglerfish. We only have about a hundred of them on camera," Sidney said, shaking his head. Sidney, appropriately named, represented Australia. In fact, he represented Australian Aborigines.

"No, we missed a video of a bathyscaphe the likes of which I've never seen." Elizabeth paused. Without the video to prove her claim, she hesitated.

"Why, what was strange about it?" asked Heather, the freckled-faced representative from the UK.

Elizabeth mentally gulped since it wasn't the caterpillar itself that caught her attention. "It was bigger than any I've ever seen, and it had a net of fish dragging behind it. I thought we were at the deepest possible depth, but this thing came from below us. When it drifted close to us, I saw the pilot, but I couldn't identify his race. He was bald and pale, almost snow white." She wasn't about to say that he didn't look human, lest they think she had a screw loose in her head.

"Wasn't Tyrone with you?" Heather continued. "Did he see it?"

"He didn't get a good look."

"My fellow distinguished team members, this is a matter for our meeting, which is to be held in five minutes," Yusuf said in a tone Elizabeth found haughty. "Our leader, Sanjay, is not present."

Sanjay and Tyrone entered the room as if hearing their names made them appear. "I'm here now," Sanjay said. "I see the meeting has started without me. Do I gather that Elizabeth spotted another scaphe? If so, that's at the top of our agenda. Did you see the scaphe?" Sanjay asked, turning to Tyrone.

"Yeah, but I didn't get a good look." Tyrone had smooth pecan-brown skin and rarely spoke up, as if his stick-out ears made his brain capable of listening only.

"Why do you assume it was a *him*?" Heather asked.

"I don't know many bald women," Tyrone said with a straight face, inviting laughter.

"Good one, Tyrone," Elizabeth said, pleased that her shy friend had responded, even if she didn't know whether his remark was intended to be amusing. She suspected that part of what he liked about piloting the caterpillar was not having to interact with a group of people. Only one or two others could fit on board the bathyscaphe. Elizabeth loved seeing the fish in the flesh she had studied for her degree in Marine Biology. Other members of the team found explorations boring after a few outings.

Command Center expected the ROE team to rotate all required activities, but Sanjay understood and respected that people's backgrounds and personalities led to different interests and allowed people to assign themselves to their preferences.

"Well, maybe he's an alien, and all of his kind are bald," Heather said, smiling. "Hair would be floating in your eyes if you lived underwater." She stroked her long, caramel-colored hair. It took Elizabeth aback that she joked about what Elizabeth wasn't willing to suggest, at least not yet.

"Good point, I guess I don't know, but for now, let's call it a he," Elizabeth said.

Before anyone else could add to the *bald* discussion, Elizabeth shook her head. "Folks, we're missing the point. Who else could have sophisticated deep ocean technology?"

"It must be a country that didn't want to join or support our mission. The Russians is my guess," said Sanjay. Silence ensued as those in the lounge contemplated his conjecture. It was true that Russia had been the only major country that refused to join the international community in creating this massively expensive station. They were also among the countries that refused to contribute despite agreeing that it was less expensive than establishing cities on Mars.

"Relax, Elizabeth," Sanjay said, shaking his pleasant face topped by thick black hair. "If someone else is down here, we'll see them again. I'll report your sighting and see if Command Center has a clue. And I apologize. I'm the one who removed the last recording, and I may have accidentally bumped the off switch in installing a new one. To be safe, let's agree that explorers check the monitors before they leave to explore."

"Sorry for my hissy fit. It's a simple mistake. It's just really unfortunate something like this didn't get recorded," Elizabeth said, calming herself, but she wouldn't rest until another team member experienced the being she saw. The men on the team had accused her of overreacting when she warned someone about doing something she considered dangerous. She couldn't help speaking up and viewed herself as appropriately cautious and someone who followed rules even when they made no sense. She thought some of her colleagues took chances they had no business taking. But Sanjay might be right, and the pilot she saw was wearing a strange body suit. However, that didn't explain that the scaphe seemed to have come from an impossible depth, which was impossible for a bathyscaphe.

Tyrone and Elizabeth both represented the US, the major donor. Other countries had a single representative. Tyrone usually teamed up with her on the explorations. He was the best navigator of the team and willingly volunteered. She feared he'd had some bad experiences as a Black growing up. It bugged her that society labeled anyone non-white as black.

Elizabeth had participated in the worldwide youth revolution and found it no surprise to learn that every member of the ROE team had also been involved. Ten years after their uprising, they had successfully pressured governments that ignored the world's worst problems: climate change, poverty, population growth, and failure to respect all human rights. Further, the active youth pointed out the strong relationship among the problems. The poor have little power, and often, the homes they could afford were in the most polluted areas.

Elizabeth considered all but one on the ROE team good friends, but would they believe her if she told them the truth about who or what she thought she saw?

Alvin Chapter 2

(The alien Elizabeth saw)

Tetrapeds, Alvin and Fitz, had been on a routine exploration of the sea above their rock ceiling. "What in ashes was that?" Alvin asked.

"What was what?" Fitz asked. Then looking around, he saw it. A vehicle in a color he'd never seen before. Both koms knew only the government of Pelagia had bathyscaphes. In fact, Pelagia essentially governed the other five small countries in their world. Other than the color, it could have been a newly designed aquascaphe, but as aquanauts, they would have known about it.

"Let's go after it," Alvin said.

Fitz frowned. "It's too late, it's already too high, and it's moving up. The difference in pressure that high would make us explode."

"Then whoever is in it can't be tetraped if it's run by beings that can survive at low pressure." Alvin's tone of voice highlighted his amazement.

Aquanauts Alvin and Fitz's routine ventures explored as far above Pelagia's ceiling as it was safe. Fitz piloted, and Alvin scanned the aquasphere for whales, edible fish, and seamounts. They composed one team of two who regularly explored. The fish above the ceiling were not as contaminated by pollution as those below in their country of Pelagia, and the uppers claimed them for their meals. It's not clear they cared about scientific discoveries as much as capturing safer fish for their consumption.

ROE Team Backgrounds

Heather's Story

Heather grew up on a sheep farm in Northern England, James Harriot's country of *All Creatures Great and Small*. She milked cows, gathered eggs, and tended livestock as well as household chores. However, the creatures that fascinated her were the butterflies, ants, aphids, moths, spiders, and anything that crawled or flew.

She became upset at learning that even insects were threatened by climate change.

When she arrived at the University of York, it surprised her that she could major in entomology. She rejected becoming a veterinarian because she had seen enough blood and gore on her family's farm. No one would call an entomologist in the middle of the night to see to an ailing army of ants.

She transferred to the University of Cambridge and received graduate degrees by earning scholarships and working as a waitress in the faculty cafeteria.

Sanjay's Story

Sanjay was born in New Delhi. His parents died in a car accident when he was eleven. His paternal grandparents sent him, their only grandson, to Westminster International boarding school in London. His two younger sisters were raised by his maternal grandparents. He suffered emotionally for a year over the loss of not only his parents but frequent contact with his siblings. As an adult, he matured faster than most; he had no choice.

While cordial, he had no close friends for over a year. Slowly, he adjusted and made friends from various countries. He developed a propensity to appreciate differences in people while believing that, at their cores, humans were much the same.

He attended Oxford College, where he became proficient in French, German, Spanish, and Farsi while majoring in International Relations. He stayed with his grandparents in London and commuted by train. After college, he was offered an internship in the US. He attributed his being chosen for the ROE team to his honest answer to a question posed by his interviewer as to what he found most interesting about different cultures. Without having to think, he said he had found that people living under repressive governments admired lying and cheating while governments that respected citizens admired honesty and programs that supported the common good.

Chapter 3

(Potential problem in the nursery)

Elizabeth rose early and pulled on one of her army dull-green uniforms, an outfit identical to the team's. Her busy brain kept her from sleeping well, and she fought against it torturing her further. Besides, she had garden duty today, which always relaxed her. The smell of freshly brewed coffee lured her into the dining hall.

As usual, Hans was on breakfast duty. "What will it be? The usual or the unusual?" he asked — his favorite joke — wearing his usual smirk. Hans represented Germany. Always upbeat, he told the corniest jokes. They were often so bad, Elizabeth found them funny. They reminded her of her dad's jokes, which fit the label "Dad Jokes" like a Phillips screwdriver fit a screw head. One of her father's favorites referred to hens thinking 'any old cockle doodle doo.'

"I'll have my usual peanut butter granola bar," she said. Besides granola bars, the food island stocked sundry boxes of cereal, bread, butter, a toaster, and a microwave. " No, wait! I'll have the unusual, stewed greens from our very own garden." It amused her to play along with Hans.

The first bite of greens almost made her gag, but she was determined to develop a taste. Now, she wished she had asked Hans for an omelet. The coffee oddly relaxed her after it hit. She noticed Hans observing her. "What?" she asked.

"You seemed spooked yesterday."

"That's because I was. How'd you feel if you encountered a bathyscaphe with an alien-looking pilot staring at you?" Oops, she had intended to avoid the use of the word *alien*. "And we don't know of any other country involved in deep-sea research," she quickly added, despite thinking the guy she saw definitely did not look Russian.

"Oh, did he have horns on his head?"

"No horns," Elizabeth said, chuckling to make Hans feel clever. "I guess the benefit is that finally, Command Center has something legitimate to do." Many, including Hans, resented how Command Center presented themselves to the world as if they were living on Yellow Submarine and in daily charge of every activity, particularly those which appealed to the public. "I won't have peace of mind until they discover who else could be exploring our territory." She picked up her coffee cup, refilled it, and said in leaving, "The garden is calling me. Our plant life may need some TLC."

She headed down to the lower level. It took a few seconds for her eyes to adjust to the bright lights of the nursery room. They were set to provide simulated sunshine early in the morning. Long tables were covered with trays with the sun lamps shining down on the little green babies. She checked the irrigation pipes; all appeared to be working; then, wearing rubber gloves, she lifted every moisture stick carefully while inspecting the plants for any signs of distress. Time passed quickly, and everything looked great until she encountered two strawberry plants in a back corner. Her heart rate went into overdrive. The leaves displayed unusual brown moldy-looking spots with white centers. Not good! She plucked them from their trays and sealed them in a glass box designed for that purpose. Everything had been going so well down here that she dreaded the implication that growing food in this environment might be problematic. She headed for the dining room, hoping to catch Pierre, who was in joint charge of the nursery.

She asked Hans if Pierre had eaten yet. Hearing no, she ordered scrambled eggs and waited. She didn't relish being the conveyor of worrying news two days in a row. She was finishing her eggs when Pierre, the tall French botany expert with a pencil mustache matching his eyebrows, came in. His eyes widened when Elizabeth informed him about the plants. Elizabeth followed as he rushed to the nursery. They snapped on rubber gloves at the entrance while she led him to the mottled wounded. He lifted the glass cover and gently pushed back the damaged leaves. Concern never left his face, but he looked up, and without a word, Elizabeth led him to the tray in question as if it were the scene of a tragic accident.

"I'm glad you didn't rearrange the others," he said as he visually inspected them, bending his head every which way. "You're sure these are the only ones?" he asked. It pleased Elizabeth that he took her nod at face value. Seemingly satisfied, Pierre picked up the cover from under the table and placed it over the guilty tray. He glanced inside the nearby wastebasket. "Those the gloves you wore?"

When Elizabeth nodded, he tossed his in, and she followed suit. He gathered the trash liner's top, pulled the built-in sash tight, and took it with him.

"And the verdict is?" Elizabeth asked as they strolled down the hall.

"Damned if I know, as you Americans like to say. I will order the on-call crew to do a sterile cleaning and institute new rules. Two people in here at a time, who will check that each other is following protocol," he spits out. "I will ensure the clinic has extra face masks."

"So, it's serious."

"Yes, I fear so. I have never seen anything like it, but it is a fungus. I will see what I can find online. Let us hope it is an aberration or an uncommon but harmless problem. This is our first potentially serious issue."

"That makes two aberrations in two days for me," Elizabeth said.

Pierre looked quizzical for a few seconds. "Oh, I see. You saw this strange person in a bathyscaphe yesterday." This time, his concern was directed at her.

"Yeah, it spooks me."

"I understand." Then he managed a tight smile. "I will have something to eat, and if you are up for it, please join me to give the nursery a thorough cleaning."

"Count me in." This was the first time Elizabeth had seen Pierre so serious. He'd dealt with mistakes in over-fertilizing or underwatering a tray here and there as expected routine mistakes. Someone once knocked over an entire tray. Of course, such accidents didn't affect the entire nursery. She remembered her dad keeping mold out of corn by letting it dry in the fields. Sometimes, the price for corn was so low that farmers waited to sell it. The federal government assisted by paying farmers to store their crops, which gave rise to the large cylindrical galvanized bins that looked like giant tin cans topped by an appropriately sized tin-man's hat from the Wizard of Oz. They could be seen from inner space dotting the Midwest. Elizabeth couldn't keep a smile from her face as she reminisced about playing in the back of trucks filled with corn. The kernels were smooth and fit between her toes. She always went barefoot around the farm, much to her mom's chagrin when a neighbor hinted that she was not a concerned and caring mother.

Elizabeth realized Pierre stood waiting for her to walk with him to the cafeteria. "You were in a dream world," he said. The dining room was deserted, and they sat with second cups of coffee. A large room on the submarine kept chickens, providing meat and eggs. Manufacturing plant-based meat and meat cultivated from animal cells would be implemented in another year if life on Yellow Submarine went well. Other staples were regularly delivered from the harbor and would be for the entire time of the experiment. Eventually, the nursery would supply a greater variety of fresh vegetables. Since young adulthood, Elizabeth stopped eating beef because of its significant impact on the environment. It was a pleasant surprise to discover that the plant burgers tasted better and more like ground beef from her dad's cows than the national grocery stores sold. She never mentioned it to her dad, however.

Pierre postponed his turn for library work and supervised Elizabeth, Alessandra, and Heather, who volunteered to help sterilize the nursery. Heather, another farmer's daughter from England, was a perfect workmate. Heather soberly understood the gravity of the situation and worked well with Elizabeth. Alessandra kept making small talk at Pierre's side while he busied himself with a close examination of each leaf.

At dinner, Elizabeth deliberately sat next to Haru, the representative from Japan, who had scouted the sea with Tyrone that day. Haru's keen, dark eyes, under a dark mop of hair, pulled to one side, studied Elizabeth and answered the question she wanted to ask. "No alien scaphes out there today."

Elizabeth said nothing, merely making a face. She imagined it would be some time before their paths crossed again, if ever.

"Oh, please understand me. I did not mean to say that what you saw was not real. I see you upset. Are you OK?"

"I'm definitely OK, excited about what I saw. I'm just afraid the rest of the team won't believe me. It could have been a once-in-a-lifetime event."

"You never know," Haru said. He was another favorite of Elizabeth's. He was one of the kindest men she had ever known. In fact, warmth emanated from his face despite his horseshoe mustache that curled down the sides of his lips.

"Do not worry; everyone respects you. We all know how levelheaded you are."

"Thanks, that makes me feel better." But it didn't.

ROE Team Backgrounds (Hans and Haru)

Hans's Story

Hans, the German representative, was named Hannah at birth. However, when Hannah was about to start Kindergarten, she told Papa and Mama her new name — Hans — and so she/he became Hans. By first grade, the older kids, who had studied English, taunted her/him, calling him 'Hans der trans.' He turned the label on its head and called *himself* Hans der Trans. Choosing sides on the playground, he insisted he be called by this name. His playmates considered it funny and accepted Hans der Trans as an OK playmate.

A chemistry set for Christmas at nine years old thrilled him. His parents warned him about experimenting with household products, explicitly about mixing bleach and ammonia, but they may as well have told him to leave the candy bars alone. He couldn't resist wanting to get a whiff of chloramine, thinking opening all the windows would make it safe, and received his only spanking ever. It was clear he needed a college with a good Chemistry department.

When his mother let him help in the kitchen, his attention turned to the chemical properties of yeast, whipped cream, egg white meringues, and the interaction of baking ingredients like baking powder and soda. His chocolate chip cookies melted in one's mouth.

He was amiable, able to laugh at himself, and his corny sense of humor made him popular in any group. He eschewed leadership but willingly served on committees. Joining Yellow Submarine appealed to his love of adventure.

Haru's Story

Haru grew up as an only child in a traditional Japanese family. His family could never buy him enough Legos. He once built an exact replica of the Kiyomizu-Dera Temple and used small flashlights he covered with pink paper to shine on his masterpiece on special days, reflecting the same pink as the real temple.

He was a quiet, bright, studious student. By age ten, he could fix any kitchen appliance, garden implement, and even routine problems with the family car's motor.

His teachers loved him, although he never raised his hand. That is, until they called on him, amazing them with his depth of knowledge and his eagerness to share more than the teacher had time to spare.

He studied Engineering at the University of Tokyo as an undergraduate, then attended UC Berkeley for graduate work in mechanical engineering.

Alvin Chapter 4

(The second encounter)

This time, Fitz first spotted the scaphe Alvin had seen. They trained their monitor towards the yellow scaphe but traveled below and behind the alien scaphe as close as they thought they could without being spotted.

When they returned to the huge garage, they eagerly played the monitor. The being sitting in front was likely the pilot, who kept his eyes on the sea in front of him, while the being behind him looked from side to side.

"I don't think we were seen," Alvin said. His face illuminated with excitement. "I bet I know the kind of being who was driving that scaphe."

"You're thinking of that time we encountered a vehicle filled with dead bodies without fins or gills."

"Yes, the ones our medical scientists studied and concluded must receive oxygen via large chest lobes. What we saw must be the living versions," Alvin said. "I want to communicate with them."

Fitz laughed. "You may as well try to talk to a shark."

"If I succeed, I'll tell you a story you won't believe."

A perplexed Fitz frowned. "If you've been holding something out on me, you'll be telling me the story anyway," He threatened.

Chapter 5

(Elizabeth spots the alien again)

Three days passed, and no further plants showed fungus. Nevertheless, Pierre had carefully sealed the diseased plants and sent them back with the next delivery vessel for analysis. Elizabeth noted that Pierre seemed constantly on edge and wondered how much time had to pass before he relaxed. She missed his wry sense of humor, enhanced by his charming French accent and pleasant nature. He seemed a different person, one who had retreated into himself.

On the fourth day, Pierre discovered two plants that looked torn. He placed them in another closed container. He wondered if someone could have accidentally damaged them. Sanjay had told him that Command Center was still looking into the fungus. Still, their reaction to Elizabeth's encounter reeked with incredulity, claiming that they would know about any such ventures by other countries. Nevertheless, they had contacted all the involved countries to alert their back channels and increase surveillance. Despite the global revolution to cooperate in stemming climate change after centuries of mistrust, remnants of the need for caution remained. Sanjay said Director Jackson acted skeptical without any film footage, which would help identify the manufacturing company of the relevant bathyscaphe. Elizabeth understood their regret all too well.

Elizabeth trusted Tyrone to check the monitor on the front of the caterpillar. Still, she became a rigid taskmaster in ensuring that anyone else taking out the caterpillar was as responsible. There was never any need to turn them off since they turned off with the caterpillar's motor, and the footage could be lifted out without turning off the monitor. She volunteered to explore more often. On the one hand, she felt desperate to see the being again, but on the other, she desperately needed corroboration of her sighting for peace of mind. Whenever she had spare time, she examined back footage. Finally, she discovered what she believed to be a portion of an oblong vehicle descending out of sight. While it was too nondescript to be of much use, it should convince Command Center there was something else man-made down there.

On Elizabeth and Tyrone's third outing since the encounter, both spotted the alien scaphe. She felt like a child waking up to a mound of presents on Christmas morning. Once they were close enough for her to gaze into the eyes of the pilot, she stared at him the way she imagined she would examine her firstborn. Oddly, she felt no fear, merely great curiosity. Who are you? Her brain tingled. At first, she attributed the sensation to her mind's inability to take in this bizarre experience. Then, she knew with certainty that the being had transmitted to her that he had come in peace and that his kind meant her no harm. The conveyed message startled her, inducing a smile from the white being. Again, without words, she received his belief that his message had reached her. She was to move her head up and down to confirm. She felt her eyes widen as she complied, causing another smile.

The alien locked eyes with her, and she thought he was asking if she could also send language-free messages with her mind. She shook her head *no*. He conveyed he understood and that nodding up and down for yes and back and forth for *no* made sense to him. Since a few countries reversed the meaning, she knew that the meaning of nods and shakes was not completely universal, but the fact that another sentient being used this same body language must mean it's a natural instinct. She had no choice but to let him lead their communication. She wanted Tyrone to share her experience and put up both palms facing the alien to indicate that he should wait. She tapped on Tyrone's shoulder. When he turned around, she pointed out her left window. When he saw the alien scaphe alongside theirs, Elizabeth looked at the white being and pointed at Tyrone. The being smiled condescendingly, verifying he was male in Elizabeth's mind. Immediately afterward, Tyrone's head swiveled like a mosquito was whining around his head.

"Did he tell you he has come in peace?" Elizabeth asked.

"How did he do that?"

"I have no clue. His kind can send messages, but I don't think they can read our minds."

"His kind?"

"I'm convinced he's not human. You'd better pay attention to the controls, so we don't bang into each other. Now you can back me up. I didn't want the rest of the team to think I've gone bonkers."

The white alien watched them, seeming to understand they were communicating. She had to contain the urge to talk to him when it was pointless, even if she could be heard. She could only turn to him, eager to experience his next message. She hoped he'd address her many questions. Calming herself, she waited.

"What I like know is your biology, technology, your culture, ordinary life, religion, customs, governance," he conveyed. "Do this," he nodded, "if it is your curiosity."

A dumbstruck Elizabeth considered carefully before nodding back. She could think of nothing else; he had hit all the vital issues she could think of. She didn't care about stupid fads that come and go. She continued to nod.

"Now I include copilot," the alien messaged to both of them. "He need not look at me." The being continued. "We live under your sea floor. At one time, we believe our ceiling the end of all there is. Ceiling is pockmarked. Some deep and some not. A curious scientist discovered that one was a tunnel with twists and turns. The end surprised us when it open to new expanse of space. Then, we explore. We find transportation vehicle with dead bodies who different from us. Big surprise. We bring them all back for scientists to study so we know about you."

"They find major difference in way oxygen go to your blood." (Elizabeth had to guess at this message.) Then, flaps opened in the alien's neck and closed again. His neck had appeared smooth before, but gentle openings and closings became apparent now. He must have exaggerated the motions for her benefit. "You have organs here," he conveyed, tapping his chest, and your kind take water from nose or mouth."

Elizabeth frowned. What transportation vehicle? Did he really mean water? Could he come from the deep sea and live in the ocean? It looks as if there were ocean water, not air, in his scaphe. Does he think the world is all water?

Are those flaps on his neck gills? Her brain swam with questions.

"You understand me?" he conveyed. His eyes widened despite having no eyebrows.

Elizabeth nodded slightly, frowning. Then she held her thumb and forefinger an inch apart, wondering if this gesture of a little bit would translate. The alien studied her. She asked Tyrone to hand her his water bottle with a lid, then poured water into the lid. She drank a little while the alien stared fixated. Then she waved her hands in the air and pretended to scoop air into her nose.

The alien looked as if he were solving a math problem in his head. Then his mouth opened, and his eyes widened. "You live in gas?"

Thrilled, Elizabeth nodded her head as if it were a bobblehead on a spring.

"Water ends and gas begins?" Incredulity took over his face.

She grinned and vigorously nodded again.

"You cannot swim in gas?"

She shook her head.

"I stunned. I need absorb before continue. There is problem with arrange meeting. I return here, but we must have different measure of time. You come back, look for me here?"

Elizabeth nodded with relief. Then it occurred to her that if they both kept track of the time-lapse until they met again, they would have a common unit from which they could arrange future meetings, but she didn't know how to suggest it. She pointed at him and then back at herself, pinched her forefingers and thumbs on each hand together, and then spread them as if she were spanning a string. Then she pointed at him and back at herself again. He looked as perplexed as a dog being given a new command, but she repeated her actions until his face brightened, and she received, "I keep time until next time, and you keep your time. Then we have a measure. It was unclear who was more thrilled with this arrangement.

Elizabeth needed time, lots of it, to digest this experience but understood the greater gap in his understanding of humans. We know fish have gills, but he did not know life outside of water. Just as she knew about space continuing outward forever, he evidently had believed water continued outward forever.

She was excited by the challenge facing her and her fellow teammates to share their world with this alien. Now, she didn't like thinking of him as an alien. The team needs to find a name for him. Alien has a negative connotation, and she deems him a wonder.

Then, an unpleasant thought hit her. Could she take him at his word that he came in peace? And why, if he lives in water, is he traveling in something like a scaphe? She grew up on a farm in the Midwest where everyone trusted each other. Now, she wondered if her lack of worry was justified.

ROE Team Backgrounds

Pierre's Story

Pierre grew up in Chartres, ninety kilometers from Paris. His parents were people of the world, who supported international causes and delayed parenthood. Once financially established, they bore three children close in age, with Pierre in the middle of two sisters. They traveled with their parents when they were not in school. If they were, grandparents from Paris came to stay with them.

Pierre received his graduate and post-graduate degrees from the Sorbonne, living with his maternal grandparents when school was in session. He majored in Biology and specialized in Botany, paying for his graduate education by teaching labs to undergraduates. He loved teaching and worked hard to make labs and subject matter interesting and understandable. He dug for unusual facts about plants, such as their ability to communicate and propensity to limit their root lengths in search of water if nearby plants are like them but otherwise spread far and wide after water. He hung a Fun Facts of the Week poster on his office door. He invited students to find more and sign their names on their contributions. His enthusiasm for his subject was infectious, and students competed to get into his classes. The techier ones would sign up online for classes ahead of their turns.

Alvin Chapter 6

(Alvin relates a secret to Fitz)

Fitz attempted several times to tell Alvin it was time to stop staring out the window at the adjacent alien scaphe. Finally, the other scaphe moved away, and Alon nodded at Fitz. "What were you doing all that time? he asked. "I can't believe you were able to communicate."

"It's a long story, one I haven't shared with you due to a promise to some tetrapeds I care about. I haven't told you the full story about the region under our sea floor."

"Well, I respect why you haven't told me the secret means of accessing that region. The fewer who live in Pelagia know, the safer the professor and his followers are. I'm glad you've told me they're doing well despite what must be primitive circumstances."

Alvin took a deep breath. "There's much more I haven't told you. Remember the mystery of the ages — the disappearances of the Blues? Not only are they living down there, but they have also thrived and developed an amazing, large city governed by fairness and justice. There are no poor. The government provides housing, food, entertainment, and transportation at no cost. Their intelligence and talents are measured, and they are given appropriate jobs with free training and education. The professor died some time ago, but his group lives among them. The Blues initially had a tough time accepting them because of their ancestor's experience with we white Pelagians." He paused to let this sink in.

"That's amazing, but it makes sense," Fitz said while nodding. "But why are you just telling me now?"

"I've only told you half of the story. Not only did they develop a new society, but they also evolved and could use their minds to communicate. This ability was critical to their survival. They needed to spread out to capture food, but it was also critical to communicate findings and call for help. These needs likely drove their evolution. Life has an uncanny way of adapting well to new conditions."

"You're not going to tell me that you know how to communicate with your mind," an incredulous Fitz studied Alvin's face.

Alvin telemessaged Fitz, "That's exactly what I was leading up to. They used some machine to prepare my brain, and then I spent time practicing. It was quite difficult."

Fitz's eyes widened so much that a portrait of him might be titled *Aghast!*

Alvin laughed and telemessaged, "Your reaction wasn't much different than the aliens we just met, even though I warned you."

"You didn't really warn me," Fitz said, his frown deepening.

Alvin laughed. "Maybe not." He continued in telemessaging mode. "The hardest thing to learn is targeting individuals when others are present and over a distance. Young Blues are taught these skills at an early age."

"I know what you are going to tell me next. I promise I will tell no one," Fitz said.

"Suffering sharkfish, you predicted correctly." Alvin patted his fins.

"Well, there's a caveat. If there's a good reason, I'll make an exception."

"Trust me, there'll never be a reason I'll accept as good." Fitz's negative reaction concerned Alvin.

ROE Team Backgrounds

Yusuf's Story

Yusuf was the only child of a wealthy family in an exclusive neighborhood of the Bourdon region of Turkey. They lived in a three-story mansion with a basement, which included a cellar, workroom, and small apartments for the staff that served his family. His parents saw no reason for him to learn how to perform routine chores. His father was a member of the Council of Ministers, and his mother, as Queen of the Household, coordinated the large staff. She held open houses four times a year, allowing the public to experience how the truly wealthy lived.

Yusuf was highly intelligent and skipped grades in an exclusive private school. He was accustomed to being the brightest and most informed person in every group he associated with. Consequently, he saw it as his calling to share his superior knowledge with his less-educated colleagues. His mother saw that he be tutored in any subject of interest to him not taught at his school. Unfortunately, he never interacted with anyone equal in intelligence and knowledge, nor many people in general. Consequently, he did not think anyone as brilliant as himself existed. He also deemed anyone with political and moral opinions different from his family's as inferior beings.

He received his bachelor's degree through PhD in Physics at Bougacici University in Istanbul and couldn't bother making friends.

Chapter 7

(Elizabeth seeks a name for the alien)

Elizabeth prepared her report of her mind-blowing encounter for the team's next meeting. So eager were her teammates for details that they bombarded her as she began speaking in the gathering room. They talked and interrupted her every word, asking question after question. Finally, Elizabeth said in exasperation, "I'm telling you everything I know. Please let me finish, and then you'll know as much as I do." When she came to the unbelievable idea that his kind had come upon a transportation vehicle containing human bodies, she said, "I must have gotten that wrong. How on Earth could that have happened?"

"Good Heavens!" said Haihong, who was the oldest member of the team. "You must be too young to remember the Malaysian plane that went down but was never found?"

The entire team went silent, their mouths agape, and their eyes widened. Even Elizabeth reacted similarly. She had indeed heard about that plane, but it was a long-ago event when she first heard about it. The proverbial light bulb flashed in her brain. "The aliens must have taken bodies from the plane and examined them carefully. That's precisely what the US would do if an alien body were found at the ocean's depths."

Elizabeth finished her report with the alien's conjectures about humans, which now made so much more sense. She ended with, "I am no longer comfortable calling him the alien. He may be foreign to us, but he is a sentient being and deserves an appropriate name. Think about it, and let's decide at our next team meeting. Any thoughts come to mind now?"

Big smiles spread across the room. Some even shouted out their first ideas: Seaman, Halibut, Yomi, Baldhead, Snoop Fish, …

Of course, Hans had to joke, "If we choose Seaman or Halibut, we'll have to be very careful about its pronunciation." Heather was sitting next to him and gave him the old elbow.

Elizabeth started to laugh for the first time in days. Then tears began streaming down her cheeks, and she could barely speak. "You guys are great, but please, no fish or silly names." She had felt so much pressure to get the team to believe her, that she hadn't been herself. Now, she and the team were on the same page again, and she felt the camaraderie she had been missing.

Chapter 8

(Pierre and Elizabeth connect)

Elizabeth felt even better hearing colleagues at breakfast tables talking about the properties the alien's name should have while she was in line for scrambled eggs. She recalled a valuable truth about human nature and group dynamics. She had been the sole provider of information about a super-duper major discovery, and this couldn't help but set her apart from her teammates. Now, something as trivial as naming the alien had brought them together. She couldn't help smiling as she scooped the eggs onto her plate. She decided not to propose a name herself and looked forward to hearing what they'd come up with.

She overheard the suggestion of names of Greek Gods. Pierre came up behind her. "You're deep in thought. I'll grab a breakfast sandwich and go to the nursery to check the plants. Want to join me?"

"Yes, see you in a bit."

Once they discovered all seemed well in the nursery except for more torn leaves, Pierre suggested a walk. Elizabeth was grateful that the submarine hadn't been designed exclusively for research, working, eating, and sleeping. The library, a movie theater — only old films — an exercise room, and a space for arts and crafts, all on the second deck, were great, but best of all was the walkway on the top floor with its view of the ocean waters. It was like deep sea diving without cumbersome rubber suits and oxygen tanks. Walking here made up for their claustrophobic bedrooms on the same level. Scientists once thought that not much life could exist this far below sea level, but they could not have been more wrong. The strangest imaginable life existed, such as the blobfish, deemed the ugliest living creature. It did look like a blob of runny grey mashed potatoes dropped on a platter. She recognized a particular anglerfish that circled Yellow Submarine because the handle sticking out over its face was peculiarly bent. She named it Bent Needle. All of the anglerfish, with their long-pointed teeth, looked unbelievably ferocious. The prongs on their heads could serve as light poles to enable hunts for food.

Pierre took a breath so deep that it caught Elizabeth's attention. They had been walking in silence, each in their thoughts. She looked up at his face. "I'm sorry I've been so paranoid and unpleasant since we discovered that fungus," he said, surprising Elizabeth.

"You had a good reason."

"Not so, or at least to the extent that I turned into a zombie. Is that the right English word?"

Elizabeth chuckled. "Yes, I think it's the word you're looking for, but it's not the word that describes who you are. You take your job very seriously, and you've cared for the plants with utmost caution."

"Thanks, but if humans are going to live in submarine cities, they have to be able to handle the problems that arise."

Elizabeth pondered. She so valued honesty; it was not in her nature to say only what would make a friend feel good. They walked in silence for a tenth of the circumference. The inner wall had numbers on the doors of the individual sleeping closets, as Elizabeth referred to them. They were intended to make it easy to find room assignments, but the numbers were also used to keep track of the number of rotations for walkers and which elevators went down to specific sites on the lower floors. Elizabeth realized how long she was taking to respond. "You're right. You did a great job handling the contaminated plants and keeping the nursery spotless, but you let it get to you, and I missed the old Pierre. But, hey, no one is perfect." She reached up and patted his back. "You seem more yourself, now. It was a good lesson that things will go wrong, and we must all live with it. It's no different up top. In fact, it's worse. Down here, we seem like a society that understands we're all in this together. We may have different ideas, but we support each other. I hope it's not just a matter of size. If humans survive, we'll have to have larger submarine cities. Sometimes, I wonder if a group of scientists are too special to represent random groups of people."

Pierre shook his head. "I am certain that with all of the crew that keeps the engines running, the library, the doctor and dentist's office, and even the janitors will be surveyed at some point. So, now, who's the worrier?"

"Me, I guess. Hey, let's enjoy the day. What's that bible verse? Sufficient unto the day is the evil thereof."

"Yes, let us. We should be celebrating your major discovery. What names do you think our team will offer?"

"I have no idea, but it seems to have engaged them."

Pierre took her hand, and they finished their walk hand in hand, saying no more. She viewed it as an example of the French's demonstrative manner, not a danger of violating Rule #9 of the ROE team handbook.

ROE Team Background

Haihong's Story

Haihong was born in Shenzhen, commonly known as China's Silicon Valley. It is the headquarters of several largest technology corporations, including Huawei, the telecommunication giant, and many small software companies. As an important international coastal city, Shenzhen hosts numerous national and international events annually, a source of pride for China.

Haihong's parents frequently reminded her about the importance of their city to China and the world. They played a game of asking their daughter at an incredibly young age, "What is China's most important city?"

With bright little eyes, Haihong would chirp, "Jenjen."

Her dad would act surprised and say, "Right you are!"

Her mom would then say, "Yes, you are so smart, but it is Shenjen."

When she got a little older, her dad would add, "And where is Shenzhen?"

"Gonedong" Haihong would pipe up and hop up and down.

"Right again, you little smart one," her dad would say, and mom would say, "Almost right! It's Gwondong."

With both parents involved in the tech industry, there was no doubt that their only child's future would be consistent with the city's reputation.

Haihong never thought twice about the general area where she would work someday. She narrowed it to IT and AI as an undergraduate and specialized in developing communication apps in graduate school.

Chapter 9

(One month earlier)

At first, all but a few timid team members thought exploring the deep sea in the caterpillar would amount to one long, exciting adventure. Indeed, there were moments of astonishment. However, once you've seen nothing new and suspect you've seen everything interesting, including some desperately ugly fish — the three hours in close quarters were found to be tedious.

Still, as a true marine biologist, Elizabeth was delighted each time she spotted a different fish.

In particular, the two Americans encountered strange fish, such as a long gelatinous eel that looked as if it had been dipped in olive oil and ready for a fry pan, a hermaphrodite viper fish you'd not want to see in your worst nightmare, deep-sea batfishes ambling over the seafloor on their arm-like fins, a deep-sea flatfish with both eyes on one side of its head, and the seriously ugly blobfish. The vicious-looking anglerfish were ubiquitous.

Elizabeth liked sharing life stories to stem the boredom. Tyrone claimed he looked forward to the next Chapter of Elizabeth's life just as he used to look forward to reruns of an old TV series about a black family called the Jeffersons. When Elizabeth frowned, evidently trying to think if she'd seen that show, Tyrone explained it was a spinoff of the famous All in the Family, which was a groundbreaking comedy. "You have to have heard of that show."

"Oh, yes, and they were neighbors of Archie and Edith," Elizabeth said. "I remember now. It was controversial and innovative to put racism on display." She watched Tyrone's stone-faced reaction in the small mirror hanging from the caterpillar's ceiling. He never spoke of encountering any racism, making her feel he had experienced such pain he couldn't talk about it. She suspected it explained why Tyrone seemed more comfortable speaking up with one person in the caterpillar than in team gatherings. Of course, many team members were outspoken, as often scientists are. She, herself, had been outspoken all of her life, especially for a female.

"Is the South side of Chicago as bad as its reputation?" Elizabeth asked during a long spate of cloudy water.

"No, some neighborhoods have high crime rates, so it depends on where you live. Hyde Park, Bridgeport, and the South Loop are deemed safe. The University of Chicago is in Hyde Park, making the surrounding area desirable since many students like to live off campus. Illinois Institute of Technology is in the Bronzeville area, known for its African American history and culture. Michael Reese Hospital is also a plus for Bronzeville."

"So, where did you live?" she asked, hoping he had not grown up in fear and poverty, as was the overall reputation of South Chicago.

"We lived in an apartment atop a high-rise near Hyde Park. My dad worked for the city and could take a bus to work until an accident kept him from working."

"Oh, no," Elizabeth said. "How old were you? What happened?"

"I was ten. My dad got hit by an old refrigerator sliding off a dump truck when he went to untangle the trash," Tyrone said with a straight face and straightened lips.

Elizabeth imagined he had told that story many times in his life. She reached forward and patted his right shoulder.

"My dad was resigned to be in a wheelchair for the rest of his life, but he was very talented and capable. Still, no one would hire him," he said with a hint of bitterness, "and my mom took an extra job to supplement my dad's disability payments. She wanted to use her salary from cleaning offices at the University to save for my education. My dad would sail around at home taking care of the house cleaning, laundry, and cooking."

It warmed Elizabeth's heart that Tyrone sounded proud of how well his dad managed. "That must have made it easier for your mom since she had two jobs," Elizabeth said, emphasizing *two*.

"Her second job was as a nanny to two preschoolers. Unusually, it included no housework. Mom had quit two jobs where the working women wanted laundry done, house cleaned, and even dinner prepared. She loved the kids at her current job and enjoyed playing with them. She was able to nap with them at their naptime. Mom thought she had it easy." His head wagged back and forth, and the reflection in the mirror of his face warmed Elizabeth's heart.

"She found time to be a taskmaster about my spending hours per day studying," Tyone continued. "It wasn't enough to finish the assignments. I had to read ahead — just in case. It paid off. I got a scholarship at the University of Chicago due to my mom's employment there."

Obviously, Tyrone's mom recognized his intelligence. She knew mothers working long hours has been the back story for many successful Blacks throughout history.

Chapter 10

(The Naming)

Those on cooking duty for special events outdid themselves for the meal before the meeting. Each team member wanted to make their country's favorite dishes. It delighted Elizabeth that the team considered the selection of a name for the alien a special event. Yet, a discussion about discovering a new form of sentient life had to top anyone's list. There weren't many events to celebrate on the submarine beyond holidays. Hans had made spaetzle — small German dumplings. Given a German father, Elizabeth grew up with them. Pierre contributed coq-au-vin and greens from the nursery, Alessandra made focaccia, and Yusuf made oven-baked rice pudding, and after dessert, he declared his dish as the best. Elizabeth just shook her head, whispering to Pierre, "The coq-au-vin was far and away better." Despite the special dinner, the naming of the alien provided the basis for the festive atmosphere.

Elizabeth called for nominations. Alessandra spoke up. I'm representing an informal committee that met this afternoon. We propose several names: Poseidon, Alon, Prometheus, Charon, Barack, and Thor. The first two are gods of the sea, with Poseidon better known than Alon. Prometheus, Thor, and Charon are also Greek gods. We all know the name Barack, but perhaps not that it's of Arabic origin. Her tone suggested her disapproval of the name of a former US president.

Elizabeth smiled, "These are great, very interesting. Are there other nominations?" When no one said anything, Elizabeth said, "Come on folks, we need at least one ordinary name."

Heather spoke up, nominating Edward, the most common name of English kings.

"So, we have seven options. Does anyone want to make an argument for one in particular?"

Haihong argued that the gods of the sea were the most appropriate. Haru pointed out that Poseidon was a cruel god. When Hans declared that the name Alon sounded like how a person from the deep South of the US would pronounce alien, the laughter caused Hans to bobble his head like a clown receiving a standing ovation.

Elizabeth's smile refused to leave her face. "How do you want to vote? Show of hands or a secret vote?" Elizabeth asked. The team clamored for a show of hands. The team's energy over the decision continued to surprise and delight Elizabeth. She called out each name. The two sea gods received the most votes, but neither had a majority. Narrowing down to these two, Elizabeth called for another show of hands, and Alon edged out Poseidon.

"Alien from South USA it is," quipped Hans, incapable of knowing when not to take a joke too far. He simply enjoyed the groans.

"I like it," Elizabeth said, and she meant it. Little did she know how close it came to Alvin, his name in his world. Now, what shall we call his kind?"

"Seamen," Hans offered. The men laughed, with Hans laughing the hardest. Elizabeth shook her head. "I reject the idea. Let's be serious."

Alessandra clasped her hand over the jewelry she wore around her neck above the top of the uniform zipper and spoke up immediately, arguing for a gender-neutral title.

"I completely agree," Heather said.

"How about Alonite?" Elizabeth said, despite her vow to let the team make all of the decisions. Fortunately, everyone liked it, and she could move on.

"Now, let me address a bigger issue. So far, our communication has been limited. It's basically a one-way street. We can only tell Alon something about us when he asks us yes or no questions. Also, we only learn about Alon's world from what he shares; we can't ask questions about his kind or their way of living."

After everyone murmured agreement, Sanjay said, "Regarding the communication problem, I've requested oversized pictures be included in our next delivery. Pictures that tell a story, like people swimming in a lake with others sitting on towels on an adjacent beach. Blue skies and, obviously, the sun, the moon, and stars. Then there are the animals: wild and domestic. We should invite Command Center's input once they are convinced the alien is real, oops. Sorry, I meant that Alon has the power to communicate telepathically. Any other thoughts?" He made eye contact with each member.

Sidney spoke up. "I think you have the right idea in pictures telling stories, but we need to create a list of what stories we want to tell."

"Yes," Elizabeth said emphatically. "This should be an ongoing discussion, but does anything immediately come to mind?"

"Pictures of the same person at different ages, from baby to elderly," Sidney suggested.

"The same woman in different activities like playing tennis, swimming, hiking, snow skiing, dancing, celebrating a wedding, etc." Alessandra offered.

"Means of transportation," Haihong added.

"The interior of a grocery store with all its goods," Hans said.

"A variety of plant life," Pierre added.

"Different living quarters to indicate different classes of people. I have a photo of my family's mansion and beautifully landscaped grounds," Yusuf said.

"That's a good idea," Elizabeth said with a puckish smile on her face. "Alon has indicated that they also have different classes and that their uppers are largely responsible for their pollution."

Yusuf frowned but said nothing. In fact, everyone went silent until Hans came up with "The interiors of our bathrooms with people wiping their butts."

Elizabeth's face displayed disgust, but everyone else laughed. Usually, she pretended to appreciate his jokes to make him feel good. Only once did she roar out loud. At one breakfast, Hans asked her about the tears rolling down her face while she read an article about a couple who provided foster care for seventeen children, including one that had been left in a box outside a fire station with a note that said, 'I love you.' Hans quipped, "They should have had a box that says, 'Take One, Leave One.'" It was the only time she laughed through tears.

Elizabeth ended the discussion by saying, "If anyone has appropriate pictures on hand, please get them to me. They need to be supersized to be seen."

Hans spoke up again. "There are seconds on the dinner." Since seconds of anything were rarely available, the team cheered and headed for the food bar. The meeting had exceeded Elizabeth's expectations. She had never been prouder of her colleagues and felt privileged to be part of history.

Sanjay took Elizabeth aside for a moment. "Thank you for what you've done to pull the team together. It's something I've been working on, but you did it in two meetings. It makes a huge difference."

"I couldn't agree more about the importance of teamwork. Today, women's basketball is thriving in the Midwest. It all began with a basketball team from the University of Iowa. Not only did they have a star player that stunned every sports enthusiast, she and the rest of the team attributed their success to the love the team members held for each other."

They raised their glasses of wine and clinked.

ORE Team Backgrounds (Sidney and Alessandra)

Sidney's Story

Sidney's birth parents named him Birrani, a traditional Aboriginal name. His adoptive family feared people would tease him about its resemblance to Biryani on Indian restaurant menus and changed his name.

Sidney knew he was adopted and was curious about his ancestry. His parents agreed on the importance of understanding his background. The use of native plants by Aboriginals most fascinated him, and in high school, he volunteered at a site that specialized in the most common of these plants. He convinced his parents to add Toothbrush Grevillea and Acacia to their yard. He argued that Grevillea grows fast and provides a habitat for native wildlife like frogs, lizards, bees, nectar-sucking birds, and butterflies, all critical for the environment. Acacia would attract witchetty grubs, an Aboriginal treat Sidney was fond of.

He received a Master's in Agriculture at Curtin University in Perth, specializing in his favorite and most interesting indigenous plants. While applying to attend the University of South Wales — Sydney — for their PhD program, he heard about the opportunity to join the ROE team of Yellow Submarine, which was practically located in his own backyard.

Alessandra's Story

Alessandra was born in Milan. Her mother was a fashion designer, and her father was in the financial industry. She was one of two daughters among four children. They got along well enough, but despite a large wardrobe of unique dresses, a special family event would inevitably coincide with the two sisters wanting to wear the same dress. Their mother would roll her eyes and rule the winner by flipping a coin.

Alessandra, with her beautiful dark eyes and hair, could have been Sophia Loren's daughter. Her mother taught her how to put on makeup and wear clothes that flattered her figure.

While they were financially comfortable, paying for room and board and tuition for four offspring would have been prohibitive. All four studied at the University of Milan and lived at home. Public transportation to the University was accessible. Milan's mix of tourists and students worldwide made it the perfect city to study sociology.

She wrote her Master's thesis on how tourism impacts the society of a city. Given the number of tourists that visit Milan every year, the citizens had grown weary of being treated like information booth attendants. They became cold and distant if preyed upon too often. She couldn't base her thesis on a single city, so she moved to New York City. While she was working on her thesis, she worked as a waitress. Since her beauty had garnered better than average tips, she graduated debt-free.

Chapter 11

(Bizarre history of Elizabeth's hometown)

The next time Elizabeth and Tyrone encountered Alon, he communicated that he had come only to tell her he would be busy for four cycles of his time — five days in Elizabeth's time measurement. Then Alon sped away. Elizabeth wondered why he hadn't said why.

When it seemed all the fish were away at a Deep-Sea picnic, Tyrone said, "It's your turn to tell me more about your background. What was the town like where you went to school?"

"It was called Dry Creek, and it has a strange history. Back in the late 1970s, a sequence of mysterious, baffling murders occurred. First, an entire family was found dead in their beds, looking peacefully asleep. The only cop in town, Ben Swanstrom, was beside himself waiting for an official autopsy. He had arranged for the family's furnace to be carefully checked and had researched other means for gas to seep into the house. The autopsies came back as death from lack of oxygen. It was as if they'd been sent to outer space for fifteen minutes and returned. When more people were found dead of no apparent cause, Ben was baffled but determined. The town still talks about the unusually thick fog that persisted for that entire fall and winter. The state police helped patrol the town, believing that the only way they'd catch the perpetrator of the murders was by discovering him in action. One state detective assisted Ben. Then he met an old native Lakota woman, Soledad, who once acted as a midwife and a healer. She used ancient Lakota prescriptions of herbs as remedies for illnesses." Elizabeth smiled proudly and said, "I'm a descendant of Soledad. This unusual woman suggested that the deaths were caused by an evil spirit inhabiting the air. Spirits inhabited everything in nature according to the beliefs of the Lakota. Desperate, in the end, Ben Swanstrom believed Soledad. She told him of an ancient means of chasing off evil spirits by burning herbs. When the townspeople assembled under the town clock for the ringing of the new year, Ben passed out small jars of sweetgrass, sage, and hemp. Shortly after the incense burning began, an out-of-season tornado rolled through the town square, and after that, the deaths stopped. Here's where the townspeople took one of two sides. One side believed the tornado and cessation of deaths were complete coincidences, the other that it was evidence of an evil spirit's anger and the power of herbs."

"What do *you* believe?" Tyrone asked.

"I believe Soledad if only because I'm a descendant of hers," she said with a funny smile. She did not tell him that she had dreams about Soledad and that if she believed in reincarnation she was a second Soledad.

"That's quite a story," Tyrone said.

"I know that's not very scientific of me, but the science of our Earth is always garnering new information so perhaps someday ..."

"You're right. Life has surprises. Who would have guessed that something like Alon could exist?" Tyrone paused in reflection.

"You must have stories about racism," Elizabeth ventured.

"It wasn't too bad when I was a kid. Well, I think my dad couldn't find a job because it was more about his race than his disability. Sometimes a clerk in a department store would try to serve a white person who stood behind my mom. My mom would say she was next, and usually, the white person would agree.

"How about your adulthood?" Elizabeth asked.

"I don't want to talk about it?" Tyrone said at warp speed.

"That bad, huh?"

"Yes, that bad. I don't think I'll ever recover, nor will I ever trust a white person again, even those who claim to be non-racist or maybe especially those who claim not to be racist."

Elizabeth's mouth dropped. "I'm sorry," was all she could say. She wondered if that included her.

Chapter 12

(Trouble on Yellow Submarine)

Elizabeth and Tyrone returned to learn that one of the janitors developed flu symptoms followed by another six victims shortly afterward. The designers of the submarine project and Command Center had worried about the density of the population in case of disease. No matter how well the air circulated, it didn't compare to Earth's surface. The doctor in charge of the medical facility quarantined the first victim in the modest sized medical room, but shortly afterward several more of the working crew developed symptoms and were quarantined in their small sleeping rooms. The medical staff visited them regularly. Among those ill, every known flu symptom attacked one or more of those affected. The variance of the symptoms and severity further puzzled the physicians. Sanjay ordered team members to isolate themselves in their rooms as much as possible but volunteer to do janitorial chores using disinfectants in their kitchen and dining rooms. Elizabeth rued that she wouldn't be able to take her daily walks on the upper deck, but she could continue explorations with Tyrone since they would be away from the submarine.

The unpredictable flu soon claimed the life of one of the oldest occupants, the librarian. He had had asthma his entire life, but his medication kept his condition under control well enough to pass muster to work on the submarine. He used to joke that librarians lived solitary lives on Earth so why not on a submarine?

Rather than send the afflicted to hospitals on Earth's surface, Command Center sent two of the best epidemiologists to the submarine.

Elizabeth and Tyrone somberly set off for their routine exploration. The usual fish seemed to be taking a break, and they knew Alon wouldn't be showing up.

Tyrone sighed. "I went in a different direction and we're seeing far fewer fish of interest. "I've been thinking about your amazing story about those murders in your town. What was its name again?"

"Dry Creek. Yes, it made *News of the Weird* and was listed as a potential example of mass hallucination." The disgust in Elizabeth's voice said all that Tyrone needed to know about her opinion of her town's dismissal by the outside world. "Dry Creekers themselves still argue today about what happened, a miracle or Soledad's version"

"Hey, there's a bioluminescent anglerfish! We've never seen one of those. I didn't think anglerfish could do that."

"I didn't either, but I can't believe how many different strains we've seen," Elizabeth said, as Tyrone quickly turned to follow it.

"I wonder if they're edible. We still haven't received a net we can attach to the caterpillar to capture fish," Tyrone shook his head.

"In my family we'd probably fight over the anglerfish handle. My dad ate chicken feet and so did us kids."

"Really, what do they taste like?" Tyrone frowned.

"I guess the tip of a chicken wing comes closest. There's not much meat on either one. I think they're a delicacy in China."

"Did you have to work harder because your dad liked to farm the old-fashioned way."

"Yup. Dad used his kids instead of weed killer to pull the worst of the weeds." She made a face and rolled her eyes. "I was the oldest of three sisters close in age and a brother born when I was in high school. So, we were the three musketeers: Me, Suzanne, and Margaret followed by Jonathon."

"The natives of the world grew their food sustainably, and agronomists agree their means are best for the environment?"

"Right, and now farmers are adopting regenerative farming, avoiding plowing and tilling. Dad did plow and till, but he was careful about the soil condition. It couldn't be muddy, but it couldn't be completely dry either. He used to joke at farmers who plowed dry land that he'd gladly have the top soil lost to the air end up in his fields."

Elizabeth saw Tyrone's upper body jiggling in a silent laugh and then said soberly. "Is it true that agricultural companies are creating seeds that grow infertile seeds or track the DNA of their seeds and make farmers pay if they use the offspring seeds?"

"Yes," Elizabeth said in a tone that shouted her disgust. "It's another reason I believe we have to get money out of politics so we can pass laws to protect the general public."

"People have been saying that for years," Tyrone said, "but I guess the money in politics is self-preserving. Did your dad have no rest from farm life? The old way sounds like more work."

"Good question. Maybe not, but he had fun teasing all of us. My earliest memory was of my dad coming into the house with a huge syringe after inoculating pigs. He'd pretend to be about to give me a shot. I ran away screaming even though I didn't really think he would do it. My mom always scolded him for doing things like that. "When I figured out how much fun my dad was having, I knew he wasn't serious."

Tyrone glanced at his mirror displaying an unreadable expression. "He sounds like a piece of work. My mom wouldn't tolerate that." Still, he laughed.

"All right, it isn't that funny," Elizabeth said, but she couldn't help laughing along with him.

"OK, we're almost back," Tyrone said.

"Good, you've probably had enough of Elizabeth on the farm?"

"No. You are more interesting than scanning for fish and seeing nothing for half hours at a time."

"Oh, I see. I'm less boring than staring at water."

"Let's hope we don't return to more people sick."

Chapter 13

(Elizabeth and Tyrone have an astonishing encounter)

Elizabeth and Tyrone set out on what they thought would be a boring excursion without a meeting with Alon.

They set out in a new direction that sea creatures evidently avoided. Elizabeth offered to tell Tyrone about her worst experience growing up. "One day, when I was around nine or ten, my dad came into the house with a baby bunny. He had run over its mother in a field, and the rest of the bunnies had scampered off. He said it was too young to survive on its own, and he asked if I wanted to take care of it. I was so excited. My dad found an eye dropper to give it milk, and we put it in a box on the porch. I tried to feed it every day, but it didn't seem to be swallowing. Then, one day, we came home from church, and it was dead. I cried and cried. I learned that life doesn't always go how you want it to."

"No, it definitely does not, but I assume it didn't scar you for life," Tyrone said.

"Of course not." She paused, thinking that Tyrone did have a life-scarring experience.

After silently traveling for a few minutes, Tyrone said, "You must have had dogs, and their life spans are short, at least compared to humans."

"Yes, we did, and I think losing the bunny did make me not get too attached to our dogs. We went through quite a few, but dogs on farms aren't pets. They're workers. They herd the cows to move from pasture to the barnyard. Then they have free time, and they love running all over, poking their noses in the barn, then in the chicken house, and through the groves where they can chase mice and squirrels. Cats are better mousers, however, and we had cats as well. They would shrink down and wait outside a mousehole."

They stopped talking and went into a kind of trance and were watching for something of interest when they encountered a beaked whale. Elizabeth had never seen one, and her wide eyes sparkled with wonder. "It's a gorgeous animal. I guess the time we spend seeing the same old, same old scenes is the price we pay for discovering things like him or her."

Tyrone agreed. "I'll follow it, but I'd better not get too close." After a few minutes, the whale began weaving its path. "I think it doesn't like us harassing him. He could decide to attack. We need to back away."

"Fine, makes sense to me. I can't wait for the crew to see the monitor."

When Tyrone turned, the beaked whale turned toward them. Tyrone turned up the caterpillar's speed as high as it would go, but the whale appeared to be picking up speed even faster. Tyrone tried to change directions as the whale had done to fend them off, but it only made things worse. "Hang on, Elizabeth, we're going to crash."

Thud! The beaked whale had gone under the caterpillar and hit it from below so hard, Tyrone lost control. They waited for another attack, wondering how much the caterpillar could take. Then miraculously, the whale wandered away. It was as if it had made his feelings known, and he was leaving them, thinking, "Bug me you will. So there!"

"We're not going in this direction tomorrow," Tyrone said.

"You can say that again," Elizabeth said.

When Tyrone repeated his comment, Elizabeth said. "That was as bad as Hans."

They returned to discover that several more members of the working crew had contracted the flu despite all the precautions being taken.

Chapter 14

(Death on Yellow Submarine)

When one of the crew with the flu died overnight, exploration became an escape from bad news. Elizabeth invited others to explore, but she may as well have offered them liverwurst and sauerkraut sandwiches. She hadn't even yet reported the incident with the beaked whale. Tyrone had inspected the caterpillar and found it had not been dented.

Off on the next exploration, the pair went with Tyrone chanting eenie meany miney mo, which direction shall we go? "We went in the direct opposite of our path to Alon yesterday. Let's go perpendicular to the two last paths. There are two choices: right or left?"

A delighted Elizabeth said, "You're fun today. Let's go." She paused, "Right."

"You got it."

"But if it's boring, are you OK with questions about any friends you had when you were young? We need to distract ourselves from this horrendous flu."

"You just reminded me," Tyrone said and took his time to answer the question. "I didn't have a best friend. One girl was in my class for three years in a row, and we studied together for tests. She came to our house once. My father was so excited, he grabbed flour and sugar containers and immediately began making chocolate chip cookies."

Elizabeth saw his smile in the mirror. "I've never seen him work so fast, but the first batch out of the oven barely made it before Astrid had to leave."

"That's a great story about your dad. I think it's great to remember our parents. I never went out for sports, did you?"

"No, and there were no parks within walking distance from where we lived. The swings behind my school were missing seats. The stairs up the slide were wobbly, and my mom told me not to climb them."

"Did your mom have bees in her bonnet beyond getting on your case about studying?" Elizabeth asked.

Tyrone brought his hand to his chin. "Well, she'd often begin sentences with "Sakes alive followed by describing behavior she found offensive, like church women she thought were hypocrites."

SPLAT! Something hit Elizabeth's window. Startled, her head flew backward. It was a blobfish. They really didn't have heads separate from their bodies, just eyes and lips built in. Their flabby bodies were a sickly pinkish gray with fat, wide noses hanging over their mouths. If it were cooked whole and put on a plate, it would spoil your appetite in no time. Blobfish were known as the ugliest living creatures alive.

Tyrone sped up and tried moving side to side to remove the poor thing, even though it couldn't have survived the crash. "Now it's your turn. What haven't you told me?"

Elizabeth huffed. "It seems I've told you everything. After scratching her brain, she said, "I remember how excited I was when I started Sunday school at our Lutheran church. I liked the stories the teacher told us about Jesus. My favorite was the one where he took a small piece of bread and somehow fed an entire crowd of people. My least favorite was his dying on the cross, and I didn't understand why God wanted him to do that. I learned I wasn't supposed to ask questions, but it didn't take long to figure out that Sunday school teachers didn't want you to ask questions when they didn't know the answers. I also noticed that sometimes, when they didn't know why God let terrible things happen, they'd say that God worked in mysterious ways. We just needed to trust him. Are you religious, Tyrone?"

"Well, like you, my mom took me dressed in my Sunday best to church every Sunday. I don't go to church anymore, but I believe a god exists. I just don't think he answers prayers, and that he's indifferent to human suffering. I also don't understand why my race puts such faith in God, when our history in the US is abominable. A righteous god shouldn't allow that."

Elizabeth became quiet for several moments. "Yet, Blacks can be so forgiving. I don't think I could accept what was done to the slaves. The hypocrisy that slaveowners provided food and shelter to their mistreated slaves, so that made everything acceptable disgusts me."

"You don't know how I despise that point of view. Getting back to Sunday school teachers, I was also told about God working in mysterious ways. My mom said we'd have a special place in heaven as if the tables will be turned between Blacks and Whites."

"I guess that's one way to live with injustice." Elizabeth decided not to continue down this sorry road. "Did you participate in Christmas programs or other church activities for kids?

"Nope, did you?"

"Yes, I loved Christmas time." Her eyes sparkled at her memories. "Every Christmas Eve, we kids put on a program. We recited bible verses or lines given to us for a skit by our Sunday school teachers. The skit always had boys in their dad's bathrobes with canes for crooks by their sides. They'd stand behind a manger with a baby doll in a blanket. My dad was a good carpenter, and he made the manger they're still using. The shepherds would say something like, 'It is cold on the hills tonight, Jacob, '" she said in a mocking, stilted way. "Once, a girl named Jody stood next to me and sing-songed her piece while she rocked back and forth holding onto and lifting the hem of her blue taffeta dress. My mom said afterward that she was glad I had not done that. My dad thought it was funny."

"I'm with your dad," Tyrone said, grinning. "Our church had programs too. I don't know why my mom didn't want me to participate. Too much homework to do to go to practice, I guess. Then, this was a frivolous activity. You know how my mom was super focused on my future. If an activity didn't help me get ahead, it wasn't worth doing." He suddenly leaned forward. "Look, there's another bioluminescent anglerfish." Tyrone slowed and approached to attempt a better look. It spun in the opposite direction and was soon out of sight.

"Your mom sounds determined that you succeed, but My Gawd, did you have any time for fun?" Elizabeth said, returning to their childhood discussion.

"Oh, yes, maybe not a lot, but my dad was a good cook, and we always talked about our days and current events at supper. I read the newspaper our neighbor shared. I read the funnies first. I'll always cherish those mealtimes, especially since my parents have passed."

"You know," Elizabeth said, "sometimes talking about something painful helps dull the pain. You may think this is a bad analogy, but one time, I slammed my thumb in the car door. I put up with excruciating pressure pain for two entire days because it was bleeding under my nail. When I finally went to the doctor, he drilled through the center of my nail and caused a small blood geyser. To my surprise, the pain left immediately. Just saying …"

"You're right, that is a bad analogy."

They remained silent for the rest of the trip, and returned to the submarine to be greeted with unwelcome news. Three more deaths, and Sanjay fell ill to the mysterious disease despite the isolation of the team. He had been feeling weak with muscle pain but hoped it would pass. Instead, he developed a sore throat, a high fever, and felt so weak the head physician kept him in the medical facility to keep an eye on him. The next few hours were critical.

Chapter 15

(The source of the dreaded disease)

It turned out that the medical experts who joined Yellow Submarine were given major assistance not from the physicians on board but from Hans. When he ran out of rice, he went to the ROE team pantry that stored non-perishable food, which was now not kept clean by the janitors. A circle of rat droppings surrounded the large burlap bags of rice, one of which had a chewed hole spilling rice onto the floor. His eyes opened wide at the sight. He wondered about the English name of the horrible disease carried by rice rats. Then he remembered Hantavirus was the same in English as in German. After sweeping up the evidence of rats and placing it in empty cans, he risked catching the virus by taking them to the medical facility to inform the medical staff. They, in turn, looked up which janitor was in charge of cleaning the pantry. The man was so young, he hadn't recognized the small oval black objects as anything but ordinary dirt or spilled coffee beans. He remembered seeing them in both pantries several weeks before the first person fell ill.

Hans thought Hantavirus had no cure, and for once, he had no urge to joke about the dire situation. How could he when Sanjay was in such bad shape? The epidemiologists gratefully received Hans's information. They told Hans that while there is no vaccine for Hantavirus nor a specific cure, early treatment is critical. Further, it's known that intubation, oxygen therapy, fluid replacement, and blood pressure medication are often effective. They had been monitoring and treating the symptoms individually, but they were excited and encouraged by knowing what they were dealing with.

Sanjay remained in critical condition, and the team held their collective breath.

When Elizabeth feared leaving the submarine to explore lest the worst happen during their absence, Tyrone convinced her that exploring would help take their minds off Sanjay. Further, if the worst did happen while they were gone, their grief would be postponed, and that wasn't such a bad thing.

This time out, the fish were not out to play once again. Tyrone asked if Elizabeth had any other farm chores.

"I had to gather eggs. My dad thought it wise to have different sources of income because the weather could ruin crops, and diseases could take out different livestock. For chickens, my dad built two rows of bins in a building where hens could climb or jump in and lay their eggs. It was like a post office with slots large enough for a grown chicken. The hens wanted to be moms, and they'd use their sharp beaks as weapons to fight off anyone who wanted their unborn babies. I wore shoes for egg gathering. I'd take one shoe off, put my hand inside the shoe, and press the hen's head to the side. With my other hand, I'd reach under the hen and take out her egg. Sometimes, hens hid their nests behind the barn. Once the eggs hatched, the hens would prance proudly with the babies scurrying behind them. I adored the fluffy, cute baby chicks." Her eyes danced envisioning the scene of them following in a line behind the hen.

"It must have been fun growing up on a farm," Tyrone said.

Elizabeth snorted. "You did lead a boring childhood." She never thought her chores were fun. They were the routine stuff of life on the farm, at least the unique old-fashioned way her dad farmed. "OK, you're a glutton for boredom. I told you my dad didn't spray weeds. Once a year, it was weed-pulling time. My dad deemed the thistles the only ones he needed to get rid of. They were tall, and their prickly balls would break open, and the wind would spread their seeds all over, stealing nutrients from the soil. I, my dad, my sisters, and the hired man would walk down rows of corn and be responsible for our row and the two adjacent ones to pull out these nasty things. Once, I encountered a really tall thistle, and I pulled so hard I landed on my butt when it finally let loose."

Now, Tyrone shook his head, smiling. "You were quite the kid. I would have given that one up or told my dad I couldn't get it."

Elizabeth made a face. "You don't know my dad." She arched her brows. "I did have some fun doing it. Once, my cousin came to help. He was in the row closest to me, and he started throwing clods of dirt at me. Then he'd run when I started throwing at him. We got carried away and lost our places. I tried to guess, but when we arrived at the end, we were way off. My dad was waiting, looking serious. To my surprise, he didn't say anything. He just pointed at our new rows and said sternly, 'Stay in your rows.'"

Tyrone smiled. "I think I'd like growing up on a farm."
Elizabeth laughed. "So, you didn't have any chores?"

"Not really. I had to keep my room clean and help with the dishes. Did you have to milk cows?"

"I didn't have to, but it looked like fun. My dad would bend a teat and squirt milk into the mouths of one of our cats. If his aim was a little off, it'd leap up into place. I asked him to teach me. It was easy. All you had to do was remember to squeeze the top of the teat and, while holding tight, pull down. You could rest your head between the cow's stomach and hind leg."

Tyrone went quiet.

"What are you thinking?" Elizabeth asked. "You seem deep in thought."

"I think it must be good for kids ' self-esteem to do adult chores growing up. It must be good for self-esteem. You don't seem to have qualms about anything."

"Interesting. I never thought about that, but I think you're right. My dad was strict and crusty. If he said to one of us kids, 'Jump,' he told us we should ask, 'How high?' Still, my dad always claimed his kids minded like fence posts. He never said 'Yes,' to anything we asked for. He'd either say, 'Hell, no' or 'What do you want to do that for?' which meant he didn't mind if we did. I thought that was odd and not like the fathers on TV."

"My dad sounds normal compared to yours," Tyrone said. "Yours sounds like something else." He shook his head.

"You got that right. He was something else. He taught us never to be afraid of a bull, even if he snorts and paws his front feet. If you are close by, you can't outrun him. You can only convince him you're not afraid, and then he's likely to leave you alone. He was right, but there are vicious bulls when that might have been the wrong thing to do."

"Wow, I'm not sure I'd like to face a bull," Tyrone said.

"I don't think I could face one today, especially if he were pawing the ground."

"It sounds as if you worked a lot outside,"

"Well, none of the chores took very long. It was cleaning up the kitchen that I didn't like. I preferred chores that you only had to do at most once a day. I like to feel I've accomplished something when it's done."

In fact, Tyrone had made her realize he was right about her enjoying her childhood. She suspected his secret pain was related to racism and worried keeping it in might not be healthy.

"You must not have had much fun if you were always studying. Did you read books for fun?"

"Sometimes, if they met her approval."

"Hey, I think you tricked me into forgetting about Sanjay."

"Guilty as charged," Tyrone said, but I like your farm stories."

Elizabeth marveled, thinking that the adage that the grass was always greener on the other side of the fence was based on a fundamental truth. She thought living in a big city must be exciting with its stores and display windows.

Sanjay was alive when they returned but still not out of danger. In fact, the doctors feared the worst, saying the next two days would be critical.

Chapter 16

(The fight to contain rats; Alon shares his personal story)

Today was the day Elizabeth expected Alon to be back, but she so worried about Sanjay that her head felt like the coils in her brain had been anesthetized. She knew she needed to accept that she couldn't do anything to help Sanjay, but she couldn't bring herself to visit Alon for days. Tyrone tried to convince her that because the doctors had ordered all of the food in the pantries be dumped when the doors opened for the caterpillar, their venture would be helping the cause of containing the Hantavirus even if it was too late to help Sanjay.

When they encountered Alon at the usual time, Elizabeth's spirits lifted. At this meet, his covering of his personal history gave Elizabeth and Tyrone a sense of family life down under. They learned he had always wanted to be an aquanaut, and his parents supported him. Once he succeeded, his fellow aquanauts claimed that he had all of the luck, and from the stories he told Elizabeth and Tyrone, they agreed he did. In addition to coming upon the transportation vehicle with dead human bodies, he and his copilot had encountered the first body of a whale aquanauts ever discovered. Every familiar fish and some Alon had never known existed teemed around the body, nibbling away. The next day, when he and his partner checked it out, they discovered a skeleton had replaced the whale's body, with only a few nibblers remaining.

As Alon continued, it became clear he relished not only the attention of the press and his fellow aquanauts but also the kind of glamour popular with the opposite gender. He seemed to both feel pride and questionable shame about taking advantage of his appeal. Elizabeth imagined him a bit of a rake at one time.

Unfortunately for him, he conveyed that the one female that attracted him most was attached to another. Landley belonged to a youth group run by Professor Thurgood from the university, who gave lectures at which he provided unsettling statistics about changes in their aquasphere. The professor formed a student organization that held both public meetings in the science auditorium and private meetings in the basement of the science building. Hoping to spend time with his new crush, Alon had joined. At the private meetings, the group made plans to convince Alon's country of the need to revolt against the government to either convince them to tend to the environment or to take over the government itself. The needed solution was clear. It was a matter of properly handling waste and appropriately filtering the water environment. The uppers blocked the solution because of the cost. They ruled the country and kept their estates filtered but weren't willing to provide safe environments for anyone else. The university in his city did find the funds to filter the campus, which became a popular place for Alonites in the surrounding area to take a swim.

While Alon conveyed with regret that he failed to win over Landley from her police officer komfriend, he continued promoting the professor's causes. In fact,

Elizabeth suspected the loss helped him grow out of his rakish ways. He risked his life in leading the project to widen and straighten the tunnel between their country and our sea floor since he had to manage a large whirling saw. Elizabeth suspected his success made him even more popular, but for some reason, she gathered he no longer relished his popularity as he once might have.

When he described the efforts of the youth group, Elizabeth pointed at herself and tried to indicate that his experience well mirrored the recent history of her society. She couldn't tell if she had succeeded. This session made her feel more connected to Alon than any other, and she speculated about the similarities between living beings on Earth. Dogs, cats, and wild animals all have two eyes, a nose, ears, four limbs, and a level of intelligence. Further, pet owners are well aware of the emotions their beloved pets experience. She had thought of sea creatures as being exceptions because of the variety of body structures. The octopus has eight legs, and each tip has a component of the brain. Marine biology did not have a satisfying explanation of how a water environment could impact the nature of life in the oceans.

She experienced indescribable feelings as she watched Alon's scaphe disappear. Tyrone didn't always participate in the communication, but on this occasion, he absorbed it all. Both considered it curious that two different sentient societies had similar problems; and in both cases, it was the youth who championed reform.

"What did you think about what we learned?" Elizabeth asked.

"He is so much like us. I wonder why he couldn't win over the Alonite female he fell for," he said. "It seems he appealed to every other female but not the one he really cared for." His face displayed his heartfelt empathy.

"Who knows? His government sounds much more repressive than ours. My gawd, they have to hold secret meetings!" She shook her head.

"I'm not sure that our government is any better," Tyrone said. "Our government and uppers try to have it both ways. Pretend we live in a democracy and that our government and corporations engage in solutions."

"You're right," Elizabeth conceded. She sighed. "OK, let's get back and see how Sanjay is doing." When she thought about all of the positive attributes of this capable kind man, she feared her thoughts too well fit a eulogy.

When Elizabeth and Tyrone returned to the submarine, Hans stood somberly waiting for them. "We're holding an emergency meeting regarding Sanjay. His condition hasn't changed, and the doctors are saying he's in crisis mode. They've tried everything except an experimental drug that hasn't been fully tested. They'd like to try it, but they need consent from a next of kin, which they are having a hard time contacting. Command Center contacted Hans. I don't remember seeing it in our contracts, but I guess the ROE team is required to give consent in a case of emergencies for something like this. Command Center wants unanimous approval from the entire team. We are holding up an emergency meeting for you two."

Elizabeth gulped and turned to Tyrone, their concerned faces speaking volumes about their fear. Surely, the vote would be unanimous.

Chapter 17

(The tense meeting)

Hans, the default leader after he discovered the rats, led the meeting. The epidemiologist attended to answer questions. Hans began by asking for a vote: "There's no point in discussion if we're all in agreement."

"I know you've told us about the experimental drug, but can we have Dr. Foster tell us in his own words?" Alessandra asked.

"Of course." Hans nodded at the doctor.

"The drug has yet to be named, but it is a nucleoside analog similar to ribavirin. It's been administered to suckling mice after the onset of clinical signs of the disease, which reduced the viral load and lethality by 63% compared to control sets of animals. Tests on human beings have been inconclusive for a number of reasons, including a clinical trial of 300 infected in China showing similar improvement to the trial with mice, and another in Russia which showed no results. The trial in China indicates that the drug is ineffective on patients whose condition has reached the cardiopulmonary phase."

"I have no clue what some of those medical terms mean, but you said it doesn't work if some phase has been reached. Has Sanjay reached that phase?" Alessandra asked.

"No, he's been treated to prevent that from occurring, but he's very close."

"Are there any other questions?" Hans asked. For once, he was serious. When no one responded, he called for a vote.

Elizabeth felt her blood come to a boil when Alessandra and only Alessandra voted *no*.

Hans's blood left his face in shock. "You're joking, right Alessandra?" He faked a laugh. Elizabeth silently groaned. She couldn't imagine a worse response.

Indeed, Alessandra straightened her spine. "I most certainly was not."

Elizabeth felt as desperate as a student caught cheating being sent to the principal. What the hell could she say to calm Alessandra and turn her vote around? She knew someone else needed to dissuade her. She turned and locked eyes with Pierre, who sat next to her. He looked puzzled. How could he not know that Alessandra fancied him? Then she realized that he didn't know how often she would stop at the nursery and leave if he wasn't there.

Everyone sat silent for an awkward period of time. Elizabeth continued to face Pierre. After he looked quizzically at her, he finally got the message. "Dr. Foster, can you tell us the downside of trying this drug?"

"In my opinion, there is no downside. As I mentioned, only in the case of the disease affecting the heart did the drug accomplish nothing," evidently abandoning medical language. "It did not hasten any end results. The FDA in America is highly respected because of its caution, but while caution prevents unintended consequences, it can prevent potential cures. Yes, the FDA has yet to approve the drug, which for your colleague is unfortunate." He smiled at Alessandra. "Are you willing to share your concern?"

Dr. Foster's smile could light an auditorium. His eyes spoke great interest in anyone he focused on, and they shone at Alessandra.

"I don't know. The idea of an FDA-unapproved drug just makes me uneasy."

"Life has no guarantees," Pierre said facing Alessandra.

Elizabeth held her breath as she saw Alessandra reconsidering, and she prayed Hans would not make some inappropriate joke. In fact, she gave him a stern look that silenced him.

It was a war of the faces. The team waited in silence.

"All right, against my better judgment, I will consider changing my vote," Alessandra said, smiling at Pierre. "But I'd like to talk with Pierre a bit more."

Dr. Foster gulped. "How much time do you need? Sanjay's condition is serious and every minute counts?"

"I don't know. It remains to be seen. I only want to hear from Pierre. We'll let you know."

"I'll be waiting near your leader's bed for your decision."

Elizabeth left the lounge immediately. The rest of the team left perplexed.

Chapter 18

(Pierre and Alessandra meet)

Alessandra suggested they retire to a back corner and Pierre agreed. She got comfortable in the corner, with a soft bench attached to an adjacent wall, and patted the space beside her. Pierre hesitated but sat down across from her.

"Can we just chat a bit while my unconscious brain thinks about the issue?" Alessandra asked. "We never have a chance to get to know each other." She smiled sweetly and touched his arm.

"Anything in particular you'd like to talk about?" Pierre decided going along provided the best chance to convince her.

"Just how we like serving on the ROE team and the list of rules. I think you do a wonderful job with the nursery."

"Thanks, I don't have time to visit the library very often, but when I do, it appears you do a great job of keeping everything in order."

"I try. Have you ever been to Milan? I bet you would love it."

Pierre strained to follow her lead. "I've been there once over a weekend for a conference." He paused. "How do you like being on the ROE team? You said something about the rules."

"Actually, I'm finding it a bit boring. I miss city life?" She twirled a strand of her hair around a forefinger. "I especially miss going out on dates to the theater, fine dining, and walks in parks. I think Rule # 9 is stupid."

Pierre said nothing, trying to think. Alessandra asked for his opinion.

"I haven't thought about it much, but I don't see anything wrong with a married couple being on the ROE team. In fact, it would make sense. I don't know how many of us left behind close relationships."

"I'm so glad you agree." Another pretty smile.

Pierre took a deep breath. "It's only sensible, and I like how sensible the women on the team are. That said, perhaps we should get to the issue of Sanjay's situation. Dr. Foster said time was of the essence."

"OK, so convince me."

He repeated what had been said at the meeting and emphasized that there was nothing to lose and everything to gain in going forward.

"OK, you have convinced me. Let's talk again soon."

"Let's both go to tell Dr. Foster. We may need to formalize this at a meeting so the decision is in the minutes, but I see no reason for the good doctor not to begin treatment."

"Yes, let's tell him."

Once given permission, Dr. Foster immediately put a liquid form of the drug into the jar above the dripline keeping Sanjay hydrated.

Chapter 19

(Three on the caterpillar)

Since the rest of the team had not been informed of Alessandra's decision, Elizabeth couldn't sleep worrying about Sanjay and wondering if she could look Alessandra in her gorgeous face ever again. She knew the woman craved attention, and while she sounded as if concern motivated her vote, Elizabeth didn't believe it. Alessandra's behavior reminded her of a three-year-old wanting to put on her ballet costume and prance for friends of her parents.

Everyone came early for breakfast to learn whether Alessandra had changed her mind. She beamed as she announced that she and Pierre had decided together that it made good sense.

Pierre had planned to join Elizabeth and Tyrone on the caterpillar, but thought better of it. Strictly speaking, the team had not made the final decision. He wanted to be around should Alessandra change her mind again. When he went to the caterpillar to inform Elizabeth and Tyrone, he did not mention the reason.

Elizabeth sighed with tears in her eyes. "Thank you for changing Alessandra's vote."

"I was surprised *you* didn't respond during the meeting."

Elizabeth shrugged. "I think the team has heard enough from me. I suspect Alessandra respects the opinions of men more than women. Not all sexists are male."

"Changing the subject, as you Americans like to say, did you connect with Alon yesterday?" Pierre asked, directing the question at Tyrone.

"Yes, he told us about his personal life," Tyrone said.

Elizabeth smiled. "Yes, give him the gist."

Tyrone gave Elizabeth a look that could worm a dog, but he well summarized Alon's background, evidently comfortable enough in an audience of two people.

"It's interesting how Alonite history is so much like ours with pollution ruining our countries," Pierre said. "Out of curiosity, how did you two become involved with fighting climate change?"

"Elizabeth first," Tyrone said.

She rolled her eyes. "It's a long story. You'd need to join us on the caterpillar."

Pierre paused, evidently considering, and agreed.

Once they were launched, Elizabeth began. "OK, I was in Miss Miller's biology class in high school when she showed a film on octopuses. I couldn't believe their intelligence. That film so interested me, I joined Miss Miller's biology club, which included discussions about the environmental problem. It convinced me that my generation would live to suffer the consequences if nothing was done. Therefore, the burden of getting the world off fossil fuels fell on us. I wrote my congressman. The response was one of those beat-around-the-bush letters. I was so furious that I recruited six friends to meet with Representative Howard to urge support for HR 91340, a good environmental bill.

To get an appointment, I told the receptionist I was writing an article for my school newspaper."

Pierre chuckled. "Very clever."

"When we arrived at his office, we had to wait for him to finish with a businessman, and then he tried to charm us with praise. When I asked about HR 91340, he looked annoyed, and said, 'The bill's author means well, and I support the general goal, but it will never pass'"

One of my friends asked, 'But you will vote for it?'

The jerk said, 'I'm thinking of proposing a less extreme bill.'

I almost shouted, 'You think the bill is extreme?' He just twisted his tie. The meeting went downhill from there, but we had filmed it and put the recording on every social media we could think of, and it went viral. Not long after, the youth movement kicked into high gear."

"I love that story," Tyrone said. "Oh, we're almost at the meeting site." When they spotted Alon's vehicle approaching, Pierre's face brightened. "I see why you thought some other country might be competing with us."

"Wait until you meet him," Elizabeth said as their vehicles crawled closer. She waved as soon as she caught sight of the now familiar white face.

Alon waved back and messaged. "I see you have someone else meet."

Elizabeth nodded and waited for Alon to begin. He addressed Pierre with a welcome, and Elizabeth laughed at his surprised face. Alon's eyes twinkled despite no eyelids.

Alon telemessaged that he had told them much about *his* society, but he did not know much about Elizabeth's beyond getting oxygen from gas and having environmental issues and government problems. He composed yes-or-no questions. "Are there many beings your kind?" Elizabeth nodded, although she didn't know the world population. "Is your ceiling high?"

Elizabeth didn't know how to respond, and Alon looked quizzically at her. She tried nodding rapidly and dropping her jaw. How could she convey that they had no ceiling? Alon tried: "Your ceiling is so high you cannot see it?" Elizabeth nodded with a big, knowing grin. Close enough.

How could she ever tell him about the sun, moon, and stars? Even showing him pictures would not provide context. He continued with questions about lifespan, about needing two genders to create a new being, and more. However, Alon was in a *you don't know what you don't know status*. His questions further made Elizabeth determined to find better means of sharing information despite the seemingly impossible task.

Chapter 20

(Sanjay's condition, Team meeting)

Elizabeth, Tyrone, and Pierre returned to bad news/good news. One of the crew who had been given the experimental drug had died, but Sanjay's condition had stabilized. More important, Dr. Foster was optimistic. He said that the death was expected as the victim had entered the cardiopulmonary phase before treatment.

The routine meetings conducted regularly in the dining lounge after their evening meals turned non-routine and more frequent. Often, there were reports and queries from Command Center to review, and tonight Elizabeth anticipated hearing that there were no other countries exploring the deep ocean. Yusuf insisted on replacing Sanjay instead of Hans in the situation room and conducted the evening meeting, arguing that the meeting had nothing to do with Hantavirus. It should be run by someone with leadership qualities.

When Haru said, with Yusuf out of the dining room, "Sometimes it is easier and quicker to just give someone the rope." Hans replied, "I'd love to use the rope on him." Everyone at the table laughed with glee.

Yusuf began the meeting by announcing Command Center's tentative acceptance that Alon was not human, then continued, "But the challenge now is convincing them that the being we discovered can communicate via telepathy."

"Does anyone know any examples of telepathy among humans or animals?" asked Heather. "I have heard claims of dogs howling at the exact moment that a son of the family has been killed in action in war, and other similar events?"

"That's an interesting question," Yusuf conceded. Everyone was quiet. "Anyone?"

"I remember hearing a long time ago about an American scientist teaching some kind of rodents how to run mazes or maybe something else," Haru said. "Then they tried teaching the same species somewhere in Europe, and they learned twice as quickly. I heard this on a reputable public news station."

Haihong spoke up. "I know scientists have tested human beings to guess at which of two boxes a red ball was under. On average, one expects close to 50% accuracy, but they discovered people who inevitably scored much higher. I know that's not telepathy, but …"

Haru added, "How many of us here have experienced instances where you felt certain about something about to happen and were correct?" Almost all of the team nodded after some thought.

Yusuf ruled that the conversation had gone off course. "Let us stick to the subject at hand, people. I want to hear more from those who have experienced Alon. Pierre?"

"What most surprised me about Alon's telemessage is that it came through in French. That's phenomenal." Pierre wagged his handsome head.

"I propose we form a planning committee on how to present our findings to Command Center," Yusuf said. "And yes, everyone will vote on the final plan," he added, frowning as if he were reluctant to concede any power to the team.

"I assume the committee will include those who have already interfaced with Alon," Elizabeth said, "and anyone interested should experience meeting with him."

"Of course," Yusuf said. "Anything else, Elizabeth?"

"The most interesting thing Alon has conveyed is how much his government is like the United States, except they're even worse. There is no pretense of democracy. Corporate leaders are formally in charge of governing Alon's country. Alonites once had a democracy, but corporations bought the politicians, and later, they dispensed with them altogether and governed directly. They are seriously suffering from pollution. Sound familiar? Alon and others are working to change things, especially curbing pollution. Since he can't mindread, we need to brainstorm about how to let him know what we did to bring about a global revolution. I did nod vigorously when he speculated about beginning with their youth. He is good at directing me in ways to respond."

"That has nothing to do with how we convince Command Center," Yusuf said curtly, "and that was my question."

"I have checked the monitor from today's visit with the alien, or rather, Alon," Yusuf corrected, "and I'm satisfied that Command Center cannot ignore the evidence of two sessions. Of course, I will be on the committee, but I am satisfied. I don't need to waste my time in Elizabeth's caterpillar," he said. "I'm sending an account of the encounter I witnessed via the monitor first thing this morning."

Elizabeth kept quiet despite her reaction to Yusuf's thinking that watching the monitor proved nothing about what was happening inside anyone's brain.

Pierre spoke up. "I wonder if someone from Command Center needs to have the experience to believe Alon's ability. I'd like to be a mouse in the corner when Command Center discusses our find."

Elizabeth looked around the room. It was hard to gauge the reactions of her teammates. On one hand, the media and governments around the world would find the existence of another civilization on their own planet beyond sensational. This, despite years of speculation about life in outer space and providing fodder for science fiction. Yet, the team's excitement exhibited a slightly subdued quality as if the discovery were sacred. Elizabeth understood.

"What about the fungus on the strawberry plants?" Pierre asked. "Have they made any progress?"

"Ah, yes. I did ask and it's looking like an alien fungus. It appears that we are encountering living things that don't exist on the surface of the Earth. I, for one, am not surprised. The analysis shows that it is similar to some species in the Zygomykota phylum. At my bidding, Command Center will send a few fungicides that kill that species with their next delivery. Let us hope that is the end of that."

Elizabeth gave Pierre a look of sympathy. He merely shrugged. Elizabeth noticed that he had begun to understand that he shouldn't bear the weight of the living lab entirely on his shoulders.

Chapter 21

(Elizabeth ponders and reminisces)

Elizabeth took the steps to the walkway to peer at the open ocean and random fish. Her steps were lightened by Sanjay's sign of small improvement, even if not out of the proverbial woods. The expanse of the deep sea and moving sea life through the display windows curbed her frequent sense of being a canned sardine. Trips in the caterpillar were indeed confining. Everywhere besides the walk on the top floor modeled efficiency at the expense of aesthetics provided by spaciousness. No matter how exhausted, she dreaded crawling into her cot at night with only a foot of clearance to the ceiling. Though adequate for her few uniforms, the tiny closet added to her mild claustrophobia. It was smaller than the childhood space she had created as her private space. Her dad, Harry, was a good mechanic and bought old vehicles to store in their grove. When a tractor needed a part, he'd find one that would work in one of the old rusting vehicles. Neighbors would call on him, and he enjoyed coming to the rescue. Elizabeth's mom had reservations about all this junk in the grove. It was OK in the summer when the leaves of the grove hid the old decaying cars and trucks, but not in the winter when the trees were bare, and she hated the junkyard look.

While she understood her mother's concern, she loved the grove. The back of an old van made a perfect spot to call her own. She took a car seat from one of the cars with a spring sticking up in one corner, covered it with an old blanket for a cot, and found an ancient orange crate for an end table. It was especially great when the hired man was single, and she lost her bedroom to him.

The grove had two mulberry trees. She loved mulberries. She and her sisters would climb up into their branches to gather them even though overripe berries on the ground stained their bare feet. After the first couple of heavy snows, she thought the grove looked like a winter wonderland with everything covered with layers of white snow.

Lost in reminiscing during her walk, she almost arrived late for the meeting of the special committee.

Chapter 22

(Conference in the Situation Room)

Yusuf, Pierre, Tyrone, and Elizabeth squeezed into the Situation Room. The room was designed for the chief officer alone to communicate with Command Center and the outside world. Yusuf had ruled that the committee only involve the four.

"But we cannot meet here," Pierre said, making a face at the close quarters where he, Tyrone, and Elizabeth stood behind Yusuf, sitting in front of the myriad of dials and switches on the control board, a virtual cockpit.

"I don't think we need four," Pierre said. "Let me be excused. The nursery is enough for me to worry about. Three should fit even if it's still a tight fit."

Elizabeth made a face of rueful agreement. "Yusuf, do you think there's any chance someone from Command Center would come along with the next delivery sub?"

"That is indeed the ideal way to convince them," Yusuf said.

"Nothing like seeing for yourself," Tyrone added.

"So, I assume you'll propose they send someone, so we don't need to meet here until we know their answer," Elizabeth said.

"We're finished when I say we're finished," Yusuf said. "We need to include accounts of encounters with Alon.

"Each encounter is in my reports," Elizabeth said, and Sanjay sent all of the ones I'd written up before he fell ill. I have yet to write up today's. Do you want me to summarize them?"

"Of course. Who knows how many members of Command Center have actually read them? Then, Tyrone, I want you and Pierre to review Elizabeth's account. Make this your top priority," he paused. "*Now,* we are done here," he said, his tone of voice once again indicating that their meetings would end only when he decreed they'd end.

Elizabeth wondered if she could trust Yusuf to make a compelling case.

Chapter 23

(Tyrone describes 'the talk')

After discussing Alon as if he were a problem instead of a scientific wonder, Elizabeth especially looked forward to seeing him.

She also worried that Tyrone's keeping his pain to himself prevented him from enjoying life, and even suspected he had applied for the ROE team to escape life in the US. Racism may have gone underground, but subtle remnants have remained. He did suspect his dad's inability to find a job related more to his color than his handicap, but he'd never said anything about his own experiences.

"We haven't seen any fish for miles," Elizabeth said to Tyrone. Let's share stories about life before Yellow Submarine. You first."

Tyrone rolled his eyes. "What do you want to know?"

Elizabeth shook her head. "You're impossible. OK, tell me, "Did one or both of your parents have 'the talk'? I understand every black mother or father has to have with her children."

"I really don't like to go there." His face seemed set.

"Another time, then?" Elizabeth ventured.

Tyrone didn't respond.

After several minutes, Tyrone claimed. "Racism doesn't really bother me. I learned to live with it. I was only five when Mom scolded me when we were grocery shopping. She had skipped the candy aisle. She was busy looking up at a top shelf for something, and I ran off to get a Reese's candy bar. When I came back and showed it to her, she grabbed it out of my hand, rushed the cart to the candy shelves, and put it back. She told me I can't do things like that, or people will think you're fixing to steal it. I didn't understand and said that she picked up things. She said, 'But I put them in the cart right away. Then they know you fixing to buy it.' I told her that other kids pick up stuff they want their moms to buy. I was confused because she told me we needed to talk when we got home.

I helped her put away the groceries, wondering when we were going to talk. Finally, she said, 'Let's sit at the kitchen table.' I climbed onto a chair and waited. 'The problem was that you be running with a candy bar. The clerks in the store would think you fixing to steal it.

I said, 'That's dumb. I wouldn't do that.' Mom was quiet for a long time before she said anything. I just waited

Finally, she said, 'Here's the thing about running with a candy bar. You a black boy, and some people think Blacks like to steal.' When I asked why, she just said she didn't know. 'They just do, you hear?'

That was only the first talk about how being Black made me different from white kids. I remember telling Mom, 'That's not fair,' and she says that life is not fair. When you're a kid, you think life is supposed to be fair, and if something that's not fair happens, someone needs to fix it."

"You've hit on a universal truth. Some things can't be fixed, but in a good society, people and government should try to fix what they can."

Both were silent until they met Alon. Elizabeth contemplated Tyrone's feelings, and she realized she had been cruel to ask Tyrone about something so sensitive, a story he probably had suppressed all of his life. There are things you don't prod your friends about.

She apologized for being so nosey. Tyrone said, "It's OK. I trust you, and we're friends. I guess I just don't like to talk about it."

Chapter 24

(Alon's history)

When they met with Alon, he looked pensive, as if he were considering what to tele-message. Elizabeth and Tyrone waited curiously. Finally, Alon began with his history. He told them the story he had told Fitz about the Blues and how the Whites discriminated against them. Elizabeth and Tyrone exchanged glances at the coincidence. Elizabeth wondered if Tyrone would have liked his Black ancestors to have escaped to another world. Who knew how great their society would be if we didn't have to deal with prejudice?

After Alon had told this amazing story, he telemessaged, "I sorry, I must leave. I keep our meeting secret. I don't trust what my government does if they know about you. They like I find them clean fish and explore, so I must be careful about time." Alon dropped out of sight.

"Boy, was that ever interesting? Tyrone said.

Elizabeth nodded. "His story should be on Netflix. Blues would be Season Two, with Alonites featured in Season One. I'm afraid if we pass this on, Command Center will think we're telling fairy tales."

"Speaking of history, it's your turn. You've never told me about your college years," Tyrone said.

"After what we just heard, it will be boring. I assume you'll tell me about yours on our next trip."

"We'll see," Tyrone said in a taunting tone.

"I'll be brief for a change," Elizabeth began. "I studied hard during the school year, but I spent most of my summers protesting in front of Congress that they weren't doing enough to stop climate disasters, let alone change. I was jailed once and hit with a billy club. I still have the scar." She pointed to a spot under the left side of her chin. I don't think we got much publicity, but the number of protesters continued to grow. Police began to escort Congress members past us. We were a major nuisance, but we were not fighting Congress so much as trying to get them to pay less attention to their donors and more to climate change. That sums it up."

Chapter 25

(Trouble in the nursery again?)

When they returned, Pierre stood waiting for them. "I wanted you to know as soon as possible that Sanjay is, how you say, out of the woods. He is to have a few more days of bed rest, but Dr. Foster is now confident he will make it."

Elizabeth couldn't help hugging him.

"I should greet you with good news more often," Pierre said with a mischievous grin.

The next morning, Elizabeth had free time, and she headed to the nursery, hoping no one else would be there. She needed time to recover from her worry over Sanjay and her dilemma over communicating life on land to Alon. Plants sounded like ideal company. Lady Luck blessed her, and she deeply inhaled the fresh smell of greens. Only a few trays needed a sip of water. Soon a smile captured her lips as she gently touched radish leaves that reminded her of volunteer plants near the silo in her childhood farmyard. The smile disappeared when a faint spot underneath one of the leaves caught her attention. The spot differed from the mysterious fungus she had found before, but it could have been in an early stage. Not wanting to take a chance, she began to lift the pot from the tray.

"What are you doing?" a voice rumbled, startling her. She hadn't heard Pierre come in. Then, in a moderate tone, he asked, "Did you find more fungus?"

"Look at this. It's not like the one we've seen before, but I thought I'd set it aside to be safe."

Pierre took it from her hand, examined it, and frowned. "I'm not sure, either."

"I'm sorry. I'm not on shift. I just needed to get away from everyone, and the nursery is one of my soothing spaces. I didn't intend to do anything with the plants."

His face relaxed. "I'm sorry for how I growled. You were bending over, and I couldn't tell it was you. Put on some gloves and help me with this. I'll take this row, indicating the one with the questionable plant."

They worked for over an hour without finding another problem. Pierre's relief radiated from his face.

"Can't dry spots form on plants for no particular reason or from uneven watering?" Elizabeth asked. "I often saw brown spots on plants in our garden on the farm."

"Yes, brown spots from the sun or too much rain are common in nature, but light conditions in this nursery are highly controlled. The sprinkler system doesn't work as well as it could. It will be interesting to watch this develop. The fungus isn't spreading on the first ones we found."

"I do wish you could stop worrying," Elizabeth said. She looked him in the eyes. "It's not your fault, and it's not all on you."

"I know, I know."

Elizabeth broke the awkward silence that followed by asking, "Would you like to go with me today in the caterpillar? I could use some help thinking of ways to communicate with Alon. The more you receive his messages directly, the more likely you may have ideas."

Pierre's eyes looked up at the ceiling, then down. "Tomorrow. I want to keep an eye on these plants for another day."

Her effort to distract Pierre hadn't worked. He obviously treated the fact that if food can't be grown on a submarine, the submarine experiment fails. He evidently felt he had a noose around his neck.

Chapter 26

(Elizabeth Muses About Long-Term Change)

Elizabeth's sharing her efforts as a youth in the climate fight with Tyrone caused her to reflect on that intense period of her life. Unfortunately, the final acceptance of the fact that humans do drive climate change along with the accompanying serious nature of the consequences hadn't been recognized in time to prevent severe droughts and floods worldwide, causing wars over access to clean water, and people attempting to immigrate away from land no longer arable. For once, however, humans responded not by developing more weapons and armies but by employing scientists to find cheaper means of desalinating and adding alkali to ocean water, searching for under-ocean aquifers, and finding ways to capture melting ice at the poles. Scientists revisited an expensive way to create photovoltaic cells that use sunlight both as a means to generate electricity and distill water at the same time. With much time, effort, and expense, they found a way to make the process foreseeably affordable in the future. On the other hand, humans could not afford *not* to pay the price.

Despite this, well-established fossil fuel energy companies would not willingly go into the good night. There was the difficulty of accomplishing a transition from fossil fuels to clean energy without disrupting access to energy in the process. Humans were unwilling to give up all the comforts and pleasures that required electricity or another energy form. Promises are easily made, but ensuring that the US Congress and their counterparts in other countries live up to their promises requires oversight. Perhaps the most difficult is finding funding for all of this while the world seemed to be involved in never-ending wars.

 The US felt obligated to defend democracy, but they didn't need another battle besides fighting climate change. In fact, Elizabeth understood that oligarchs cared only about themselves and their families as if wealth could protect them from the consequences of inaction. The rich in every country were buying land with groundwater far from polluting factories and rising oceans as if that would allow them to live as always, unaware of Earth's rotation and winds carrying pollutants to every corner of the world. As Elizabeth mused about recent history, she wondered about the possibility of communicating with Alon about how hard it had been to achieve their limited successes. When Alon described his governance, which had devolved from democracy, she couldn't help but think how close her civilization had come to following the same path. His job was harder by tenfold because of a government with no pretense of democracy. On the other hand, they were a single country with a government so powerful that wars never happened. Non-patriots who were discovered disappeared. She gathered that suspected traitors were eliminated when Alon told her about a group of concerned citizens escaping to a secret region underneath their floor to avoid being assassinated.

Her one big regret about her political participation and graduate education meant spending little time with her parents. She thought she had plenty of time, but her father died in a car accident shortly after she received her master's degree, and a few years later, her mother began to show signs of early dementia. She lost all sense of time, waking up from a Saturday afternoon nap, getting ready for Sunday services, and then sitting in the dark outside her church, wondering why no one else was there. She left faucets running and a basement refrigerator unplugged, substituted doctor for minister in her speech, and misaddressed envelopes to pay bills. Elizabeth assumed the heartbreaking job of getting her into a memory care facility. Her mother resisted every step of the way, claiming she could care for herself. She did have a moment of lucidity when she said, "I'm not with myself anymore."

Joining Yellow Submarine meant she'd be out of touch for nearly a year, but her mother no longer recognized her, and difficult visits left Elizabeth miserable. She convinced herself that her mother was in a good place with many friends and neighbors looking in. Elizabeth's sisters were a few hours away and managed regular visits. Her mother loved her grandkids, and her condition didn't change her enjoyment of their visits, providing Elizabeth some comfort.

Chapter 27

(Alon's unusual request)

When the plants' brown spots hadn't spread by the next morning, Pierre agreed to join Elizabeth and Tyrone. Elizabeth found it strange that no one else wanted to experience Alon's ability to telemessage. She knew people didn't like claustrophobia, but the more members of the team who experienced mind messaging, the better for many reasons. In particular, she believed interacting with Alon would prompt more ideas for ways of describing human life to him and remove any doubts of Command Center about Alon's ability. Pierre wasn't ready to pilot from just one session watching Tyrone, but Elizabeth would love it if he could become proficient. For reasons she didn't understand, she hadn't wanted to pilot.

Today, Alon conveyed he wanted her to calculate the length of her scaphe as a unit of measurement to allow him to learn how many such units higher her home was from their current position. She held up her hand for him to wait while she checked. She asked Tyrone if he remembered the length of the caterpillar, and to check Yellow Submarine's depth and calculate it in terms of caterpillar lengths. She asked for it in the closest power of eight. Alon has four fingers on each hand, and she wildly hoped they might express numbers in base eight. When Tyrone frowned, she said. "You're the mathematician. You should love it."

"I don't love arithmetic."

"Then, use your damned calculator."

Pierre laughed at the exchange while Tyrone rolled his eyes, and Alon telemessaged that he didn't understand what was funny. Pierre waved his hands to his side, palms up, hoping that would signal he didn't really know.

Once Elizabeth had the number, she struggled to convey the appropriate power of eight. Eventually, she succeeded only because Alonites, fortunately, did have a base numbering system with eight as the base.

"That is lucky, "Pierre said. "Especially when not all countries in our world express numbers using a base system."

"But it is the best system for calculations. Imagine how many more kids would hate arithmetic if they had to multiply using Roman numerals," Tyrone said. "What's this all about?"

"I don't have a clue why Alon wants to know." She shook her head.

They didn't have to wait long to learn the answer. He telemessaged, "Can I follow you partway back to your home? I like to see."

Now, Elizabeth wondered whether he meant her home on Earth or on Yellow Submarine. He wouldn't know they were living in a structure in the ocean.

While Tyrone explained to Pierre how complicated Roman numerals made calculations, Elizabeth said, "Quiet, you two. I think Alon has more questions, and I can't understand him when you are talking."

They quieted, and Alon continued telemessaging both Elizabeth and Pierre with questions: "Do you eat food other than fish and plants? Do you have good healthcare? Do you have different classes of people? Do they have different privileges? Do you have public transportation? Do many have their own transportation vehicles? How many beings are there in your world?"

Elizabeth signaled accordingly to each one. Pierre didn't try to participate beyond absorbing the messages.

"That is all I want know now," Alon messaged finally. She became more determined to find a way to initiate conveying information to Alon.

Elizabeth called out for Tyrone to return to the station slowly. She and Pierre watched the rear monitors to ensure Alon followed. Elizabeth asked Tyrone to stop after she saw Alon had stopped. Then she was shocked to be telemessaged from quite a distance that Alon's scaphe could go no higher, but he could see the location. She wondered how far Alonites could telemessage. After docking, she told Yusuf to broadcast that people might be able to see the Alonite scaphe from the walkway. He refused. She told Hans to spread the word, then headed there herself. It amused her to see Alon's scaphe's lights circling below.

Chapter 28

(Fight over leadership of ROE)

Elizabeth turned a cartwheel on the walk deck when she learned the next morning that Sanjay was declared free of the Hantavirus. However, the doctors advised that he take a break from leading the team. Sanjay disagreed, saying he felt fine, needed just a couple more days of rest, and that getting back to his role was precisely what he needed. Further, he argued that three of the team were already going on their one-month breaks because Pierre postponed his turn over his concern for the nursery. However, Yusuf volunteered to shorten his break and take over the leadership afterward. His offer at a meeting led by Hans met with silence. Elizabeth mentally groaned until she realized she wouldn't have to suffer his leadership, nor would Pierre. Sanjay had conveyed his objection, insisting he would be OK.

Yusuf's fellow team members were conflicted. They didn't look forward to Yusuf's leadership nor wanted Sanjay to risk his health but felt sympathetic to Sanjay's wishes. The vote to accept Yusuf's offer or allow Sanjay to return was postponed for a day. The team could not think or talk about anything else until the meeting began. Hans argued that having led meetings in Sanjay's absence, he understood that the challenge of leading and communicating with Command Center was not as easy as it might appear. Yusuf argued by expounding on the value of new ideas and his outstanding leadership ability. He cited his parents' many organizations where they held high positions.

Yusuf moved that Sanjay not be allowed to vote by proxy, obviously aware that the doctors wouldn't allow him to leave his hospital bed just yet. When Elizabeth argued that if he can't decide to continue his leadership, he should at least be allowed to vote. "People very often cannot judge their own ability," Yusuf said, delivering Sanjay a double whammy. Elizabeth mouthed under her breath, "Aren't you clever."

The motion that Yusuf should serve as the leader after handling his business barely passed after exhaustive discussion. While Elizabeth voted against it, she understood the difficulty of deciding what was in Sanjay's best interests.

Hans asked the committee working on convincing Command Center of Alon's telepathic ability to meet in a corner of the library.

"I have bad news and ambiguous good news," Hans began.

"Bad news first," Pierre said.

"I've heard from Command Center. They are furious that you alerted the alien to the submarine's location, and we have to move as soon as possible. They agreed to help us locate a new site by sending us the location of known permanent currents."

"That's my fault," Elizabeth said.

Tyrone and Pierre objected as they had gone along. "It doesn't bother me either if Alon knows," Hans said, "but we have a different perspective. They are more wary."

"So, what's the ambiguous good news?" Elizabeth asked.

"They agreed to send someone from Command Center with the next delivery. That's good news if it's to experience Alon, but if it's to give us grief, that's a different matter."

They stood in silence for a moment.

"I guess all we can do is wait," Elizabeth said, shrugging.

Hans mimicked Elizabeth's shrug. "No need for the committee to meet before they arrive."

"But you'll inform the team when you know?" Elizabeth said.

Hans grinned impishly. "You'll just have to wait and see."

After two serious meetings, Hans was back to his clownish ways.

Elizabeth couldn't sleep, imagining that this would be the one time when Alon didn't show up.

Chapter 29

(Elizabeth tries to share information with Alon)

The next delivery bathyscaphe was needed as a spare emergency vehicle since locating a new site would go faster using two caterpillars. The deliverers were not pleased to have to hang out in Yellow Submarine with nothing to do since the team didn't want people unfamiliar with Yellow Submarine to be involved with packing up and buttoning down. Further, they were being kept from the things that they needed or wanted to do in their own lives. Team ROE hid their lack of sympathy.

The site needed to be out of the way of regular currents and shark habitats. Under the circumstances, chats with Alon were to be suspended. Elizabeth feared the move meant she'd never interact with him again. Hans allowed Elizabeth one quick visit with Alon to try to communicate what was about to happen. Under the philosophy that what you don't know doesn't hurt you, Hans didn't plan to let Central Command know and asked the team to keep it quiet. Yusuf said at the relevant meeting that he did not approve. "OK, teammates, let's put it to a vote. All in favor of never telling Command Center, that Elizabeth and pilot will attempt one more contact, raise your hands." Only Yusuf voted against it, shaking his head as if his teammates were lesser beings. Elizabeth worried that he'd find a way to contact Command Center about the decision, but hopefully it would be too late.

When Elizabeth encountered Alon, he telemessaged his surprise regarding the size of Yellow Submarine. She thought he might have expected a harbor from which their caterpillar launched. "You live there?" he asked. Elizabeth nodded. She didn't know how to tell him living there was part of an experiment and imagined he might be confused since he knew their atmosphere wasn't water. He did understand that like his society *her kind* had serious pollution issues. She realized more than ever of the importance of being able to transfer information to Alon.

She cupped her right-hand palm down intending it to be Yellow Submarine, pointed at herself with her left forefinger, brought her forefinger to her cupped palm, then moved her palm around it to attempt to indicate her temporary home could move.

Alon looked puzzled, and Elizabeth repeated what the demonstration. He frowned while Elizabeth waited for him to ask something. His face turned up just as humans sometimes do when thinking. His mouth dropped and he telemessaged, "you want to tell me something about where you live?"

Elizabeth felt as excited as a child on a baseball team making it to first base for the first time. While she was trying to think of other hand signals, Alon queried "You live in two places?" Sometimes in that oversized bathyscaphe and sometimes in a gas?"

Elizabeth felt thrilled and relieved at the same time, but it was Alon's intuition not her silly puppet show that succeeded. She gave up on communicating that Yellow

Submarine was going to move but assumed that because he referred to it as a bathyscaphe, he would not be surprised about their move. The effort to communicate had taken so long, they had to turn back lest they run short of oxygen.

"You have tears in your eyes," Tyrone said while he set the caterpillar in motion.

"I don't know if we'll ever see Alon again." While she hoped the new location would not be too far from the meeting place with him, she had a bad feeling about it.

Chapter 30

(Elizabeth reminisces)

Extra nursery shifts and more walks along the outer pathway occupied Elizabeth's time, while worry occupied her mind. Command Center, once a resource, now loomed as a threat. It didn't make sense, but she couldn't help feeling something ominous waiting around the corner. Nightmares haunted her sleep. During the day, she repeatedly told herself two logical truths: moving the station was no big deal, and only idiots could fail to appreciate the significance of another intelligent species — one with communication ability beyond humans. Command Center should understand that Alonites need not be feared nor interfered with.

Her mother used to complain about her daughter's impatience. She was right. Elizabeth always admitted that patience was her weakest attribute. The closest she came to being patient was stubbornness. If she really wanted to accomplish something, no matter how difficult, she'd hang in there longer than 90% of the population who would sensibly move on. When her family moved into their new house, her new bed was one that they inherited at the recent death of her fraternal grandmother. Her twelve-year-old self did not like the big iron headboard consisting of hollow, thick iron posts and asked her dad to cut them off. When he snorted at the idea and said he was too busy, she found a steel saw and began to saw and saw and saw saw saw saw. She would get tired, do something else, and even wait until the next day, but she slowly made it halfway through each post. One day, when she was sawing, her father came into the bedroom, grabbed the saw, and finished the job in short order. She danced a little jig.

Early on, she learned that her dad's crusty exterior disguised a caring heart. The neighbors often called on his mechanical ability when they had trouble with their machinery. He was never too busy to help. He had built the cross for their Lutheran church and a Lutheran Church Camp, fixed things that went wrong at the church, and kept the pastor's home in good repair.

Her mom was as gentle and sweet as her dad was crusty. With her mom's spelling and grammar, she could have taught English. However, her sweetness made her gullible. She believed the pitches of sleazy salespeople. It should have been a federal crime to sell something to her.

My mom was as gentle and sweet as my dad was crusty. With my mom's spelling and grammar, she could have taught English. She was so trustworthy of other people she couldn't imagine a traveling salesman would stretch the truth. She proudly purchased gifts for her family from companies that sold silver-plated copies of a new bill, coins called collector items because they had clearly been imprinted after being minted, some so poorly you could cut your fingers on them. My father never believed a salesman's words, and my mother couldn't imagine anyone lying. It ought to be a federal crime to sell anything to her.

Elizabeth frequently thought about her parents in adulthood, now that they were no longer a presence in her life. These memories helped distract her from her concerns over Alon.

Pierre came into the nursery, but Elizabeth was too lost in her thoughts to notice.

"Where have you been? I haven't seen you for days," Pierre said. "Anything the matter?" He looked concerned.

"No," she lied. She didn't want to talk about her despondence regarding Alon.

"Are you sure?"

"Yes. I'll just be glad to have our visit from Command Center behind us. You must be looking forward to seeing what they have for the fungus," she said in an effort to change the subject. She had hoped no one else had noticed her crazy melancholy. She forced a smile.

He looked at her curiously. "Soon, you can forget it all when we are on our leaves. Any plans?"

"Visit my mother in a memory care facility near my hometown. My father has passed. Take long hikes in woods along a stream. I miss nature. That's about it. How about you?" Pierre had managed to distract her from her negativity, and she imagined herself among tall trees with green, green leaves and eating a larger variety of fresh fruits and vegetables.

Pierre nodded in approval at Elizabeth's agenda. "I plan to tour your country. My parents are volunteering in a township in South Africa. You can advise me on what to see."

"Just tell me what you like to do, but not now. I have a meeting with Hans." She hoped Hans would share his thoughts about Command Center's intentions.

Chapter 31

(Command Center Surprise)

Command Center approved the coordinates of the new site found by Haihong and Haru. Haru's intuitive understanding of how things are put together had made it easy for him to pick up on piloting, but only half heartedly. He'd rather repair than operate the caterpillar. In his youthful days with Legos, he'd build little cars and trucks but never played with them, merely moved on to his next construction.

A Command Center representative was expected to arrive within the hour. The team's morale continued at a low point, and mixed feelings about a Command Center visitor caused frowning faces. Elizabeth was not alone in her unease. Right or wrong about her fear, she'd soon know.

Hans invited Elizabeth to wait with him in the situation room for the signal to open the larger of two docking systems for the arrival of their special visitor. Delivery vehicles consisted of two compartments, one for human passengers and one with a bin that slid out on tracks. The one carrying the supplies would be exchanged with an empty bin, enabling the crew and team members to unload at leisure while the delivery vehicle could return without delay — not needed for this delivery. She drew a deep breath when she heard the signal.

It took forty-five minutes for the docking, including the opening and closing of different compartments and the time it took to flush the water. When Director Donald Jackson, himself, greeted them, he praised the choice of their new location and displayed excitement about Elizabeth's discovery of a fascinating new species. He patted her on the back. He wore a well-fitted navy-blue uniform with brass buttons, polished to a gleam. His small-billed cap sat smartly on his head. "I'm most anxious to meet this alien you've reported so much about. Call me Don." He presented as pleasantly friendly. Elizabeth chided herself after two weeks of nail biting. After meeting with Hans and Elizabeth, Director Jackson took a half hour to settle into his room. He had brought plant-based steaks for a festive dinner for the team and everyone else who kept Yellow Submarine running.

Hans used the intercom to call for a meeting in the lounge at 5:00 for a celebration of two years of exploration under Command Center's direction. Elizabeth knew Hans didn't expect the director himself to visit. She certainly didn't.

She looked forward to a festive event. Everyone was allowed to dump their khaki green uniforms and dress for special occasions. Having the director of Command Center on board was as special as it could get. Included with the limited amount of clothing you were allowed to bring, Elizabeth had a blue velvet scoop-necked dress that fit snugly at the top and draped beautifully around her ankles. She wore a silver choker and no other jewelry. She took time with makeup, with which she usually didn't bother and headed for the dining hall. She was fashionably late, but not by design. When Pierre saw her, he said "Oo la la." Elizabeth didn't know how to react. She noticed second looks of other coworkers.

Alessandra said, "Now, that's a dress my mother would wear."

That's quite a compliment." Elizabeth said. Alessandra often spoke of her mother's reputation as one of the finest dress designers in Milan.

Alessandra smirked, and given the emphasis on the word *mother* Elizabeth wondered if she had misunderstood the implication her team mate intended. She remembered Alessandra's stories about competition with her sisters over their elegant dresses and who looked better in them.

The pilot and copilot had toted the supplies for the special event separate from the main compartment of the delivery vehicle and laid it out with the help of Heather, Sidney, and Pierre. Elizabeth had never seen such a spread of canapes, cheeses, fruit platters, lightly roasted vegetables, and mystery tidbits along with fine wine and other beverages. Heather, who was assigned cooking duty that day, whispered into Elizabeth's ear that she was relieved by the virtual feast. When she learned that the director was visiting, she claimed she had nothing she deemed special enough to serve. She was a farmer's daughter cook, not a Hans.

Hans appeared impressed with the spread. He walked over next to Elizabeth. "Much better than I could have done." When she told him he always came up with something out of nothing, which was more impressive, he replied, "You're right. I just wanted an excuse to peek down your sexy dress." She gave him the old elbow move.

Tablecloths and candles added atmosphere and delighted a team unaccustomed to elegance. Even Don's long speech, filled with clichés, did nothing to dampen the festive mood.

Things were going well in Elizabeth's judgment. She slept better that night than she had in the last ten days and awoke refreshed. She spent the morning with Don — as he had asked her to call him — giving him a tour of Yellow Submarine. He had seen the blueprints and had toured it only once before it was occupied, but he seemed to appreciate seeing it decked out with appropriate furniture, equipment, and whatever suited the room's purpose. She answered all of his questions and began to relax, wondering why she had these apprehensive feelings about his visit. He even invited her to join him in addressing the team with questions about Alon after their excursion. This was a bit peculiar. If he had read her reports, he would know that all of the questions about encounters with Alon had been asked and answered. She bit her tongue for once but warned him they would be crowded in the caterpillar. When he looked confused, she said, "Sorry, caterpillar is my name for a bathyscaphe. That's what they look like to me." She rejected the idea of taking the larger delivery vehicle, explaining that Alon would not recognize it and might turn away.

They set off shortly after lunch to arrive at their customary time. She didn't think Alon would mind being introduced to yet another human; he seemed to enjoy surprising other people. They arrived at the usual meeting place ahead of Alon. Usually, Alon beat them to the site. After ten minutes, Elizabeth's worry began to mount, and after another ten minutes, her heart felt like it was sinking, and her nerves were struggling to tow it into place. They continued to wait. Elizabeth couldn't help looking at her watch every ten seconds while previously jovial Don glowered. After another eight minutes, Don had had it. "I'm not waiting any longer." Elizabeth felt all her blood had risen to her brains and swirled there faster and faster, driving idea after idea, including her fear Don would conclude the entire alien story was trumped up. Tyrone was setting the controls to depart when she spotted Alon's scaphe. "Tyrone, wait! He's here." She tipped her head back, and her blood descended to their proper vessels. Don, however, did not appear pleased, more like he had been put upon and that a director should never have been kept waiting.

Once they could see one another, Don stared at Alon as if he were a ghost. Elizabeth wondered if he'd even watched the video footage. When Alon telemessaged "I sorry late," she smiled and hoped splaying her hands palms up alongside her head would be interpreted as it's OK, but when he targeted a welcome message to Don, Don jumped, and nearly hit his head, which was spinning looking in every direction. Elizabeth had to suppress a laugh at his reaction. Alon did laugh but telemessaged "Sorry" again. Then, he proceeded to his topic of the day and telemessaged, "My kind like hear nice sounds, like see stories with people moving. You nod if you like." His descriptions weren't as clear as glass, but Elizabeth deduced they had music and something like movies and/or television. He typically chose one topic per meeting. He had described their buildings, which she agreed seemed similar to those of humans, except the outer walls were made of rock. Today's topic evidently was entertainment.

Don wanted to ask him questions as if Elizabeth knew how. "I can't ask him. He can't read minds."

"How do you know?"

"Because he has telemessaged me that he can't." She struggled to keep sarcasm from her voice. She glanced at Alon, amused at their quarreling.

"Sorry, I must go," Alon said.

Elizabeth nodded and called to Tyrone, "He's finished. It was short today, probably because he was late."

"Follow him," Don ordered.

"No, don't." Elizabeth nearly screamed.

"I'm in charge here. He followed you to Yellow Submarine, remember?"

Tyrone's eyes turned from Elizabeth and proceeded to obey the director. Alon was some distance ahead. A relieved Elizabeth noticed Tyrone moving slowly.

"Can't you go any faster?" the director roared.

"Doing the best, I can, sir." He managed to stay a reasonable distance behind Alon while keeping him in sight. A short time later, they could see the sea floor and not long after, the back of Alon's bathyscaphe sinking into it.

"What the hell?" the director muttered.

"You know from my reports, they live under the sea floor," Elizabeth said, deliberately as a dig that he hadn't done his homework.

"Follow it!"

"Sir, we are dangerously close to the limit of how low this bathyscaphe can go," Tyrone said. "Every thirty-three feet down, the pressure increases by one atmosphere or nearly fifteen pounds per square inch. Just another few feet, and the caterpillar will implode."

Tyrone used a necessary turn to slow down.

The director swore a blue streak while Elizabeth hid her delight.

"I can't believe you don't seem interested in seeing their country," Director Jackson said. Elizabeth no longer thought of him as *Don*.

"Alon's government is repressive, and he recently conveyed he wasn't going to tell them about meeting us. If we *could* follow him down there and they found out Alon led us there, who knows what would happen even if our caterpillar survived?"

"It's not your call."

"It's impossible, now, anyway. Tyrone, let's head back."

Jackson huffed. "At least we have the coordinates. We need to capture one of them for research."

Elizabeth gasped, shivered, and gave Tyrone a look. Her instinct that feared Director Donald Jackson's visit had proved to be correct after all.

Alvin Chapter 32

(Reaction to Jackson's chase)

"Speed up," Alvin cried to Fitz. "I think they're following us."

Fitz threw their scaphe into max speed. "Do you think it's because of the new guy we met?"

"Probably. I don't know why, but I don't like it."

"It's because there's no reason for them to follow us. You said you've told them we live under their floor and that a fan blocks the opening."

"Yes, hit the fan's controls. We want it lowered before we get there." Alvin felt puzzled. He wondered if this new being might be someone in authority over Elizabeth. If Elizabeth's kind were like ours, the authority should not be trusted. Is there something about power that turns one evil? Fitz, we need to stop exploring for a while."

"No argument from me, but what excuse can we give?"

"Let's claim our bathyscaphe was acting up, and that we want a thorough checkup before we explore again," Alvin said. "Can you tamper with the engine so that they'll find some problem? You're the engineer."

"Sure, but what about the other team of explorers?"

"Ashes, I didn't think about that. Wait, they explore in a different territory, so they should be safe."

Fitz shrugged. "You're the boss. You always are."

Alvin ignored the dig. Almost all of his fellow aquanauts resented him for his fame.

Over the next several deks, a conflicted Alvin pondered over his responsibility to his country. A formal group working to reform his government no longer existed since too many of his cohorts now lived in Abyssal — below their ground. The too few who remained in Pelagia did not have the capacity to lead a viable effort to influence the governing corporation or even an advocacy organization for justice and a better environment. However, if the gas beings posed a threat to their country of Pelagia, he felt obligated to do something. However, if the Pelagian leaders overreacted, the consequences could be dire. We don't want to get into a war. At best, the ability to learn more about an entirely new society of sentient beings could be squelched. The scientist in him believed this would be an unfortunate lost opportunity to learn about the greater environment in which they lived, one where the water stops, and gas begins.

He needed to contemplate before taking any action.

Chapter 33

(Director Jackson threatens trouble)

Elizabeth instinctively understood arguing with Jackson would get nowhere now that he had dropped his glossy veneer. They did not speak during the return, and Elizabeth headed for the circular walkway to think. Not long afterward, Tyrone joined her. Fortunately, no one else was walking. Elizabeth made a face of absolute disgust.

"I know," Tyrone said. "If it's any consolation, I deleted the coordinates where Alon disappeared." He displayed an evil grin, and Elizabeth joined him.

"Joy, oh joy, pure unadulterated joy! Tyrone, you're awesome. What about the coordinates of our meeting place?"

"I didn't delete them. Do you want me to?"

"Yes, copy them first, but if Alon is as astute as I think he is, he won't be back." Her heart turned from ice cold to warm and healthy. Not only had Tyrone likely saved Alon from disaster, she saw an expression of pleasure on his face at her joy. She hugged him.

They stopped talking when they met Alessandra walking the other way. They exchanged greetings, and Alessandra so glowingly described how much she looked forward to Elizabeth's reports on Alon, she sounded phony.

"Her makeup is always perfect," Elizabeth said, making Tyrone laugh.

"What?!"

"I don't know. Here we are talking seriously, and you're talking about makeup."

A sheepish Elizabeth shrugged. "I guess that is funny. I've just not been myself for days because I was worried about Jackson's visit." She should be able to relax now that the worst is over because the fear of a tragedy is often worse than the tragedy itself. However, something was stopping her. Something ominous.

"You have good instincts," Tyrone said. "But, you understand even if they know the exact coordinates of Alon's opening, they can't get there without inventing a stronger caterpillar."

"Yeah. I also wonder what Jackson's background is. Is he ignorant of what happens when fish have been brought up from the deep? They not only die, but their bodies are destroyed. They blow up. We need to find someone at Command Center who gets it and appeals to them. I have to talk to Hans. I wish I could talk to Sanjay, but it's probably too soon to burden him. I still like keeping Jackson from getting the coordinates."

"Who's going to tell the director the coordinates are missing."

"Do you mind? It will be more convincing if I act as if I don't know. You could promise to get them and then get back to him with a 'I don't know what happened.' You should be less suspect than I am."

"It will be fun to jerk that jackass around."

Elizabeth hugged him again. "I think I'll see if Hans is available. I'll not bother Sanjay. He needs to build his strength."

Hans was nowhere to be seen. As she approached the tiny situation room, she heard Jackson's angry voice. She spun around and headed for the nearest stairwell to go to her bedroom and waited for an hour. She became claustrophobic, couldn't nap, and was starving. She left for the dining area, picked up an egg salad sandwich, and sat in the adjacent lounge.

"There you are," Hans said as he strode into the room. "I hear your trip did not go so well." He made a curious silly face, evidently thinking his understatement was amusing.

"To say the least. Did Jackson tell you he wants to capture one of the aliens for research?" Her face could have wormed a dog.

"Yes, I have orders to send him coordinates," Hans said becoming serious. "He doesn't trust us to make the capture. They're leaving right after breakfast tomorrow. Then we can dance the light fandango."

"What do you think of that idea?"

"I like my idea; do you want to be my fandango partner?"

"I meant capturing Alon," Elizabeth said with disgust. Hans didn't know when to stop with the humor.

"Don't have a cow. I knew what you meant." Hans tapped her on the shoulder.

"I'm sorry, but I'm not in the mood for fun right now."

Hans sobered up. "Sorry. The man is a fool, but he also thinks he gets to make the calls."

"Did you have any idea he was planning this?"

"No, I think he just decided when he realized Alon's ability to telemessage was real. The jerk believes that if we pop open his head, we'll see a little lightbulb we can replicate, and voila, humans can now telemessage."

This time, Elizabeth found herself dismayed but tickled by Han's confirmation of what she feared.

Her head drooped, and Hans said nothing until she got a better hold of herself. Sometimes he had enough sense not to play the clown.

A well of anger balled up in Elizabeth's intestines, then swam to her brain, raising her feisty nature. Hans appeared startled. A defeated face had lowered, and a determined face rose up.

Chapter 34

(Elizabeth begins to plot)

Elizabeth skipped the evening meal but checked with Sanjay's doctors. "Is he up for visitors yet?"

"Let me check. He's in a common outpatient bedroom. You're Elizabeth, right?"

Elizabeth nodded, and the doctor left. When he returned, he waved his hand at Sanjay's closed door. Only one other patient, who was sound asleep, shared his room.

Sanjay's eyes lit up at the sight of her. "I've never been so thrilled to see one of those dull green jumpsuits."

"I hope you'll soon be seeing a lot of them."

"Not soon enough in my mind."

Elizabeth took his hand. "You had us so worried," she said, shaking her head with her eyes beaming pleasure at seeing him recovering. "Be sure you take it easy. After being bedridden so long, it may take some time to be able to simply walk around."

"Oh, I've been walking several times a day with a nurse I don't need."

"Really? That's great. The team is looking forward to having you back." She refrained from saying more, especially with Yusuf running the show. Sanjay, appropriately, would disapprove. His role included maintaining harmony and teamwork.

"Anything significant happen while I've been laid up?" His tone indicated he expected a 'no,' but he thrilled her by opening the opportunity. "As a matter of fact, yes. Jackson joined Tyrone and me on the caterpillar to see Alon for himself. He had the same reaction as everyone else experiencing contact by telepathy. Alon was late, and Jackson almost missed him. Then Alon needed to leave early, and Jackson insisted we follow him. Tyrone drove slow enough to lose him, probably just before Alon passed through their ceiling. Now Jackson wants to capture him or one of his kind to study." Elizabeth kept her feelings out of her voice. She had been told not to get Sanjay excited.

"What!?" Cool head Sanjay almost shouted. "That's crazy."

"I know. Remember, we have no bathyscaphes that can go as deep as Alon's ceiling, so it would take a ton of money and time to accomplish such a foolish goal."

"Now, I'm particularly bothered that I'm deemed not well enough to lead."

Elizabeth smiled. "So am I for many reasons, one being that Yusuf agrees with Jackson, but what would be great is if you do go on leave, do some digging on the cost, and what the rest of Command Center thinks. Then I can use my leave to try to convince those involved that it's a ghost hunt."

Sanjay grinned the smile of the Grinch as if thinking it was an awful, horrible, wonderful idea. Elizabeth hugged him. "I'm not supposed to stay long. I'm pleased with your reaction. I know you'll be a big help."

Elizabeth slept fitfully that night. Finally, at four a.m., she got up, went to the dining hall, and poked around for something to eat. After wolfing down a couple of granola bars, she returned to her room for a couple more hours of rest, hoping she'd sleep better when Jackson and company left. If all went well with Sanjay, he'd be returning with Jackson, which would be a perfect opportunity to express his opinions.

On library duty the next day, she shelved books as if she were one of the Walking Dead. By dinner, she had transformed from zombie Elizabeth to nosey Elizabeth and wondered how Hans would report on Jackson's intentions. Jackson had conveniently helped himself in the kitchen earlier and gone to bed.

After dinner, Hans began the meeting almost in a singsong tone, describing the director's desire to capture Alon for research; Elizabeth clenched her teeth and rubbed her hands up and down her thighs. She had thought he agreed Jackson's utterly barbarian plan had to be stopped. The room grew silent and then erupted like a baby sucking in a deep breath before wailing after being hurt.

"You must be kidding!" "What the hell? This guy's a scientist?" Many jaws dropped from shaking heads.

"You reacted as I thought and hoped you would," Hans said with a laugh as if he had just pulled off a joke on the team. "May I assume I have your vote to proceed to block this action in any way we can?"

Again, the team erupted, this time with cries of support, but a scowl. However, he said nothing for a change.

As she savored her colleague's reaction, Elizabeth's head tingled the way it did when she first met Alon. Tyrone sat beside her and asked, "What are you smiling about?"

"Alon isn't coming back. Somehow, he can not only telemessage, he can do it over a distance."

"That's amazing," Tyrone said.

"Yes, but I'm not putting it in a report to Command Center. They'll be even more eager to capture him."

Chapter 35

(Elizabeth's concern for Alon grows)

Elizabeth told Hans she hoped Alon's brief message conveying he'd not be returning meant he understood any Alonite was in jeopardy.

"If he's not returning, why do you have your panties in such a bundle? It sounds to me as if Alon gets it. I doubt he'll ever leave his country again, and he will warn his fellow Alonites not to do so. Jackson doesn't have a lone ant's chance of building a skyscraper of accomplishing his goal."

Elizabeth smiled but said, "Aren't there companies working on constructing bathyscaphes that can safely operate near the ocean floor?"

Hans sighed. "I think so, but …" He shook his head as he rolled his eyes.

Over the next day and a half, Elizabeth and Hans composed a missive for Sanjay to appeal to Director Jackson about the immorality of sacrificing a sentient being in the name of science. Further, removing Alon from his high-pressure environment would destroy not only him but any chance of learning anything about his kind.

Given Elizabeth's lack of patience, she spent every available moment of the next two days pacing the wraparound walkway at double speed, barely acknowledging other walkers. Tyrone told her that people had been asking him about her state of mind. "What did you tell them?"

"You were worried about Command Center's plans regarding Alon."

"They are not?"

"You are the one who has become Alon's friend, so I guess it's less personal to them."

Alvin Chapter 36

(Alvin makes a most unpleasant discovery)

    Alvin discovered a message on his telethon to report to Benkon Kuda, the Chair of Linophryne, the governing corporation of his water world. His eyes bulged at the idea that Benkon had become suspicious of his explorations. He hadn't seen Benkon since he was asked to give aquascaphe rides at the ceremony of the opening of the giant fan. He took that as an honor and that Benkon respected him as the most popular aquanaut, and the one who engineered the straightening of the tunnel to the upper sea. Under the circumstances, however, the summons was ominous. He drank a glass of fermented seaweed and set out.

    Alvin couldn't remember the last time he'd been in the Linophryne building. It had been built into a seamount with a large magnificent ground floor with original stone walls. No matter how many times he'd been there, the high ceilings and the mammoth amount of space couldn't help but put him in awe. He stopped at a kiosk to announce his business. He received a pass he hung around his neck, stood up straight determined not to show any anxiety, and swam up to the appropriate floor despite lifts.

    He greeted Benkon's secretary with a smile and could tell she remembered him. She rose from her desk and swam-walked to the inner office door, and peered in. "Alvin is here."

Benkon must have waved for him to come in as the secretary said, "he's ready for you."

Alvin smiled at her again and entered the room hoping he looked relaxed. "You wanted to see me?" he said.

After an exchange of pleasantries, efforts to remember how long ago Benkon presided over the ceremony at the opening of the giant fan, Benkon said, "You're probably wondering why you're here." Then he proceeded to tell Alvin that he heard something rather preposterous about him. It included the country of the Blues, his ability to telemessage, and meeting a live being that lived above them in a scaphe and that there was concern that these beings could present a challenge to Pelagia.

Alvin could tell that Benkon was skeptical, and he himself was shocked. Fitz was the only possible source. He shook his head with wide eyes. "You're joking or did you have a strange dream?"

"No, one of your colleagues brought this story to me." Benkon turned serious.

"Oh dear," Alvin said sadly while shaking his head gently."

"I always explore with Fitz as the pilot. I've noticed he hasn't been himself lately, but I didn't realize he was hallucinating. Alvin called on his innate ability to charm, which he hadn't employed since he met an attractive kem in a bar. "He must have been stung by a dream fish."

"So there's nothing to this?" Benkon asked.

"What do you think?" Alvin replied laughing.

"Well, I'm glad it amused you, but I felt obligated to follow through."

"Please don't be too hard on him. Can you see if he can be given appropriate care and time to recover?" Alvin said.

"I'll do that, but he's out as an aquanaut" Benkon began shuffling kelp documents on his desk.

"Thanks, I see you're busy." Alvin left feeling he'd been in a whirlpool. He couldn't believe what Fitz had done, but it was his bad for sharing too much. He genuinely rued having thrown Fitz into the mouth of a whale. He had little choice.

He was fortunate that the truth was so bizarre, it was easy to get Benkon to dismiss it, also lucky that he had established a relation with Benkon as the top aquanaut. He'd use the need to choose a new pilot to temporarily suspend explorations. He hoped Haggard would agree to pilot with him.

Chapter 37

(Sanjay and Jackson debate)

On the trip from Yellow Submarine to Freemantle Harbour Jackson expressed his pleasure over Sanjay's recovery. "Your team was quite bothered. You are well respected."

"We have an exceptional team. Command Center chose wisely," Sanjay said, returning a compliment." He paused and took a deep breath. "I understand you experienced Alon's ability to telemessage."

"Yes, I admit I didn't believe it. It is beyond amazing. Imagine if select humans could do that. It would not only be an easy way to communicate but would be a huge advantage in case of war. World War II was won by breaking Germany's code. The ability to communicate secretly is an invaluable weapon against aggressive countries." Jackson's enthusiasm oozed from his pores and took Sanjay aback.

Sanjay thought quickly. "But if we are able to achieve the skill, what's to keep an aggressive country from learning it?"

"We'd find a way," Jackson said. "That's not the only reason I want to understand their ability. It is because I am a true scientist. I want to know everything possible about their society. A big problem is that Elizabeth is certain Alon will never appear again. I'm afraid I'm responsible for that.

That means we have to build bathyscaphes to stand much greater pressure than they can currently. That will be pricey."

"Does that mean you'll settle for all that Elizabeth has learned?" Sanjay asked. "I'm certain she will write a thorough report. She is very concerned about any attempt to examine an Alonite's brain."

"No, I will not settle. It means we need to design a bathyscaphe capable of studying the brains of these aliens in their own environment."

"Wouldn't that be horribly expensive in time and money? I'm not sure it's even possible."

"Yes, I will need to lobby Congress for funds. If the US donates part of it, I'm hoping the other countries will join the cause."

"I see. Do you think every member of Command Center will agree?"

"They usually respect my ideas. Obviously, I haven't broached it with them, but I will soon. If we study the aliens in a space safe for them, how can they object? I asked for the coordinates just above or close to where Alon descended to his country."

"I know that the ROE team, myself included, does not like the idea," Sanjay said being unusually blunt. "This Alon is like a human being. Would you consider doing that to a human who had been born with the same ability?" Since talking with Elizabeth, he had given his argument considerable thought.

Jackson gulped. "It's not the same thing. These things are not humans. We have to agree to disagree and discontinue this discussion."

"You're right, but let me put on the record that Alonites are sentient beings, not *things*."

The two men pretended the other did not exist for the rest of the trip.

Chapter 38

(Elizabeth frets)

The more Elizabeth reflected on the feasibility of Jackson's ambitious plans, the more relieved she became. Even if he could locate the funding, it would take a considerable amount of time to accomplish the requisite equipped bathyscaphe with special medical equipment and expertise. While on her break, she'd polish her political boots to start a campaign against the director's diabolical plan. Until then, she'd work on a detailed article relating Alon's story. All the news agencies should be eager to publish it and compete for interviews. She loved the idea; she envisioned them writing opinion pieces that would cause people to march in the streets. She vowed to search the caterpillar's camera footage for a good picture of Alon to copy onto signs LEAVE ALON ALON(E)! If she lost her position on Yellow Submarine because of her opposition, so be it.

She abandoned the walkway for a computer in the library during the time between shifts, eating, and sleeping. Composing the article consumed all of her spare time before her break. Pierre read her missive and agreed it met Elizabeth's two goals: create sympathy for the aliens, and emphasized that taxpayers had better projects to fund in this time of climate crisis. Given the subject matter, he thought it couldn't help but fascinate readers.

Elizabeth remembered the last time, the awful time with Jackson aboard, she had seen fish in Alon's net trailing his scaphe. It reminded her that her recommendation for Yellow Submarine's caterpillar to equip their caterpillar had not received a response. Once she had written her missive, she suggested to Hans that they get to work, put together nets themselves, and fashion a way to attach them to the scaphe to capture fish for eating. Then, she laughed at her thoughts, which included one of her father's speech habits. According to him, neither he nor anyone else ever just did something. They went to work and did it. She took pleasure in the thought that this was a sign her father approved of her idea.

The team cut back on the number of exploration missions as they now rarely encountered species other than the usual anglerfish, batfish, white crabs, occasional blobfish or shark, and of course, no Alon. Once, when they stopped waiting for Alon, an anglerfish had swum to her window and stared at her. It radiated evil with its long sharp teeth, and she stuck out her tongue at it. She sighed. Fish sightings now took her attention during caterpillar explorations, but she couldn't help scanning the seascape for Alon from time to time. Once, she and Tyrone spent nearly an hour at the old coordinates to no avail. She justified it by noting they saw as many sea creatures as when they were moving. Her feelings about not seeing him were mixed, but relief over his safety dominated over sentiment. The fan Alon described blocking the entrance created a major issue for any effort to capture an Alonite. They'd need to tunnel through at a different spot, and she suspected or rather hoped that Alonites had chosen a spot where their ceiling was the least dense.

With only two days remaining before her month-long break, she talked to Pierre about helping her determine where members of Command Center stood regarding Jackson's plan. Something Jackson said made her believe he took for granted that the other members of Command Center would be on board with his idea. She hoped he was wrong.

"I'd be glad to help," Pierre said. "The director is how you say 'out of touch.' I cannot imagine anyone approving."

"I hope you're right, but when it comes to US politics, I've learned not to take anything for granted."

"France is not different."

Something about Pierre's face caused her to wonder. Then it hit her. So absorbed in her fear for the Alonites, she had not once asked about Pierre's fungus concerns.

"What about the fungicide Jackson brought?" she asked in an apologetic tone. "Is it working?"

He shrugged. "I wish I could be certain. It's stopped growth on the infected plants, but the fungus's growth had been slowing down so it's difficult to tell if it's due to the plants' natural ability or the fungicide. I have determined that the ones you found later were suffering from something else. There have been zero new infections," he said indicating his new optimism. I've told Hans to have more of the same fungicide sent as a preventive measure. I'll spray them personally before I leave."

"That sounds great." Best of all, Pierre sounded like himself after his weeks of worry. More likely he had come to accept that threats like this were something he had to live with, rather than entertain the possibility that it wasn't possible to grow large amounts of food on submarines.

Chapter 39

(Elizabeth plots before her departure)

Yusuf decided to leave early to take care of his business. Elizabeth convinced Hans to have a meeting before he returned to take over. While all but Yusuf opposed Jackson's plan, Elizabeth feared that when Yusuf returned to lead the team, he would try to bulldoze support or, even worse, claim support that did not exist. Everyone else had expressed opposition to the plan at prior meetings, so most were puzzled that the subject had come up again.

Once everyone had assembled, Hans began. "Elizabeth believes we need to be prepared for Yusuf to pressure us to go along with Jackson's plan. I'll let her explain."

In case Yusuf attempts to change our minds, I think it would behoove us to make a list of reasons why it doesn't make sense, and send the list to Command Center. While I believe you agree that opening the brain of a sentient being is inhumane, we need to be prepared to oppose the idea that a procedure can be concocted that would be humane. Let's also make a list of why the entire project is not feasible.

Haru spoke up for a change. "I doubt that Congress and foundations would grant enough money. In fact, I'd bet my entire monthly wage on it." Everyone laughed given the size of their pay. Since everything they could need on the submarine was taken care of, their salaries were barely enough to send ecards to their relatives through the situation room.

"But what can we do to oppose Jackson from down here?" asked Haihong.

"That's a quadrillion-dollar question," Hans clowned. I'm not sure how we can throw vinegar into the spaetzle. We've only talked about the impact on the aliens, but there's also the risk of a battle in the deep sea, where humans would be outnumbered."

"That's an excellent point,' Elizabeth said. "We don't know the population of Alon's country."

Haihong stood to be recognized. "What kind of bathyscaphe would you have to build to not only stand the pressure at Alonite's level, but have a special compartment to operate on the Alonite, which would have to have ocean water at Alon's pressure? Then wouldn't the operation have to be done by robots?"

"Haru, you're the only engineer here. What do you think?" Hans asked.

"Yes, it would need robots or arm-like contraptions that can be controlled from the other room. But, first, you also have to either find the opening Alon uses or drill down through our ocean floor to gain entrance to the Alonite world. That would be the easy part. You'd have to anesthetize the patient and assume the Alonite brain is similar to ours for the amount of anesthesia that wouldn't kill the patient. Then, what are the brain swirls going to tell you about its telemessaging ability? It all sounds like a wish-upon-a-star project. I have no idea what the required specs are for each aspect. It's simply crazy."

Everyone fell silent, including Elizabeth. Thoughts were spinning through her brain like clothes in a washing machine in its final cycle. By the time the spinning stopped, she had changed her plan to ignore the director when he met the group for a debriefing and a small reception. She needed to find out his exact intentions, even though it required interacting with a man she didn't respect. She wasn't good at hiding her contempt from the contemptuous. However, she had done it before, and she could do it again.

"Yellow Submarine to Elizabeth," Pierre said. "You look as if you've already left on your break."

"I'm just realizing how much I personally need to do to stop Jackson. The arguments everyone made are overwhelmingly convincing. I hope you use them to convince Yusuf when he gets back. You've given me more ammunition I can use on Jackson and Command Center. Thanks."

The discussion provoked a worse fear in Elizabeth. Jackson wasn't stupid. He might not give a damn whether the Alonite victim survived, and be willing to accept a flawed plan, then express regret over the destruction of an Alonite.

Chapter 40

(Elizabeth sets out on her leave)

The time remaining before Elizabeth's departure evaporated like a drop of water on an Arizona sidewalk. When Elizabeth heard the delivery vehicle would arrive soon, she, Pierre, and Hans stood waiting. Three team members gone at once wasn't desirable, but needs must as the Brits like to say.

Yusuf first climbed out of the vehicle and barely acknowledged his about-to-depart team members. Fortunately, Elizabeth felt confident that he couldn't change the team's mind about Jackson's goals. With luck, Sanjay could resume leadership by the end of her break.

Before they could board, Yusuf began grumbling. "I don't know why they couldn't have spent a little more to make us comfortable. We are among the world's elite. And then I was gone long enough to become accustomed to my family's servants doing my laundry and gourmet meals, morning, noon, and night. Now I have to habituate back to crude quarters and inferior meals."

"What, you don't enjoy eating my goulash two nights a week?" Hans said, pretending to be hurt as they stood waiting for the pilot to allow them to board.

"You do well," Yusuf said. "Don't get me wrong, but you don't have access to the finest ingredients that my family does. I'm looking forward to writing a paper on what I just discovered. You know I've published five papers in the most prestigious journals in my area."

"I think you've mentioned that before," Elizabeth said. I think everyone here has published their work in highly reputable journals, so you're in good company."

"How many do you think have published as many as five?" Yusuf said with a frown.

"Probably more than you imagine."

"Well, I plan to check on that."

The pilot called for them to board. Elizabeth wasn't tall, but she had to duck to get into the cabin portion of the scaphe. For once, Elizabeth sympathized with Yusuf. There was little elbow room, and once seated. Hans joked that everyone should cross their legs in the same direction and signal for everyone to recross at the same time if someone needed to shift positions.

Elizabeth appreciated the hanging straps as a comfortable place for her arm. "The first chance I get, I'm going for a hike in the largest open space I can find."

The others nodded at the idea.

The journey would take only an hour and a half before they'd pop up from the ocean, boated to Perth for a medical exam, enjoy an overnight rest, and then board a flight on a military chartered plane to New York City and a limousine to Command Center headquarters in the UN building. Even so, having to sit still in a more crowded environment than the caterpillar caused her the heebie-jeebies, especially as she mentally practiced communicating with Jackson.

Pierre patted her on the shoulder. "Doing all right?"

She managed a small nod and smile, appreciating his concern.

Chapter 41

(Interlude in Perth before traveling to NYC)

Once on the surface, they were transferred to a sizable boat with a bar and comfortable accommodations below deck. Drinks were free, and all three team members imbibed like parched desert camels. It's one thing to be trapped in a large submarine underwater, but traveling like pieces of cargo in a compartment of the delivery scaphe with no windows not only made it claustrophobic. Visions of being trapped because of engine failure unnerved Elizabeth.

They arrived at the hospital for checkups around 2:00 p.m. and the hotel around 5:30. Elizabeth's blood pressure measured slightly high, but otherwise, she was in good shape. When Pierre and Hans made reservations at different restaurants around 6:00, Elizabeth deferred. "I need a long walk first. Then I'll decide on what I'm hungry for."

"That sounds good," Pierre said. "I'll join you."

The concierge overheard them and called, "Don't forget your masks."

"Are we still wearing those?" Elizabeth asked. "I thought by now the air would be clean enough that we'd not need them."

"Better to be cautious, especially when you're used to filtered air, and for a lengthy period outdoors."

"Do you know the latest air quality reading?" Pierre asked.

"Only that the UN says we are close to not needing masks. Many people don't bother."

"We heard that nearly a year ago," Elizabeth and Pierre chimed.

"When you're in the hotel business, you want guests to be extra cautious."

Elizabeth looked at Pierre, who grinned. "Shall we?"

"We're going to pass," Elizabeth said. She couldn't imagine wearing those masks, which made people's noses look like pig snouts.

"Very well. Just sign the refusal form."

Elizabeth signed with a flourish and a smile on her face. Pierre watched her and chuckled, then followed suit. They left the lobby like two children playing hooky.

They followed a path down to the beach behind the hotel. Neither had appropriate shoes, and in the spirit of the moment, went barefoot. Pierre took her hand, and they trod in the sand, mesmerized by the ocean, the wide-open space, and the sky filled with fluffy clouds. Signs stating the water was unsafe to wade in caught Elizabeth's attention. She had imagined the significant effort to clean up the world, including the oceans, would be showing more positive results by this time. On the other hand, governments didn't get serious until disasters snowballed, and only then finally stopped treating each hurricane, flood, and fire, as if it were just another unexpected tragedy. A responsible company invented compostable plastics, and countries with coastal waters have removed much of the plastic from their shores. Developed countries significantly reduced the amount of pollution escaping into the air but had not found an efficient way to remove the remaining amount. Undeveloped countries had not contributed to water and air pollution but suffered along with the rest and were finally compensated. Elizabeth noted that most of the beachcombers wore masks. After about a mile, they reached a residential area and turned back.

"What are you planning to say to Jackson at our meeting?" Pierre asked.

Elizabeth sighed. "The probability of spending a lot of money and learning nothing."

"I suspect that is the case, but I wonder why he doesn't get that he's living in a dream world."

"I'm hoping to learn more about why he doesn't see what we see at our debrief. Thanks for agreeing to help."

"Of course, and now let us forget this for the rest of the evening," he said, turning her face towards his while making heart-melting eye contact. Elizabeth forced herself to look away.

They made their way back, located their shoes, found a bench behind the hotel, and brushed off most of the clinging wet sand. They dined at the place Pierre had noticed earlier, and Elizabeth succeeded in finishing her gigantic fresh salad in excellent company. The dinner provided a welcome treat to end the day.

Chapter 42

(Arrival at the hotel near the UN building in NYC)

After a night of tossing and turning with a scratchy throat, she joined Pierre and Hans on a stroll along the harbor at Perth before brunch. They found it therapeutic to gossip about Yusuf and his claim that his country had been civilized for much longer than most accounted for its superiority. "He should be called You Suffer," Elizabeth said, making Hans laugh.

"Good one," he said.

Pierre simply shook his head.

"We all appreciate our backgrounds," Elizabeth said. "I was a country bumpkin, but I knew the value of hard work, and scientific discoveries require hard work."

"You Suffer prefers enough wealth to hire people to do the hard menial tasks of life," Hans said in a voice mocking Yusuf's.

Elizabeth sighed and wished she could ignore the fools and the pompous, which Pierre said was the best way to deal with those who made one cringe. They passed several appealing restaurants, but feeling as free as seagulls, they decided to check out more. Just as Elizabeth's hunger could no longer be ignored, they came to Bathers Beach House at the Harbor. "Stop, let's read the menu in the window." After scanning the list, she announced, "Looks good to me."

Hans peeked inside. "It's not up to You Suffer's standards, but I guess it'll do."

Pierre shook his head. "All right, you two. Have you had your fun yet?" Not waiting for an answer, he continued. "Yes, let us dine," he said in a haughty voice, surprising and tickling Elizabeth and Hans.

Elizabeth understood that enough was enough, but it had felt good to let loose. Sanjay wanted crew members to get along, so they stemmed their temptations to diss You Suffer.

Elizabeth and Pierre ordered mussels, and Hans, the foodie, ordered barramundi, an Asian sea bass, commenting that it was unique to the area. He made a thumbs-up gesture at the first bite. "That's high praise coming from you," Elizabeth said.

Not enough time to do any more than a bit of window shopping before they boarded their flight, Elizabeth vowed to visit Perth's highly praised science museum for children on her next break. Seeing the little ones' excitement with the interactive exhibits would be a bonus. She had never acclimated to New York City, even after months of training there. Part of her was still the little girl who lived three-fourths of a mile from her nearest neighbor. In NYC, she encountered both astonishingly rude and incredibly helpful people and never knew which would be which, or perhaps which would be witch was a better spelling. There were signs in stores near the subway that would not give change. She understood that but was shocked when she made a purchase with the exact change and then realized she should have saved the change for the subway. The clerk refused when she offered a dollar bill and asked for her change back.

On the other side of the human spectrum, during training she planned to visit a colleague in the hospital who loved fruit and found hospital food inedible. She stopped at a fruit stand for berries and apples. When the owner learned her intent, he threw in an orange and a banana. Her musings about New York City made the flight go quickly.

Upon disembarking, a man holding a sign with WELCOME AQUANAUTS greeted them outside JFK's security zone and led them to the van. The driver whizzed through heavy traffic using the principle that his larger vehicle took precedence, and he who flinches, loses. Elizabeth held the handle over her seat so tightly, her fingernails left marks. She could never drive in this city.

Once at the office complex containing Command Center's offices, they were led to a wood-paneled room with a row of chairs behind a U-shaped table for the members and three facing chairs — a blank form on each chair. A table of drinks and inviting appetizers stood along the wall. Jackson stood waiting for a pour of wine. Elizabeth took a deep breath and casually walked to the table, picked up a plate, and stabbed a few shrimp and crackers spread with cheese and pate'. "Look at all the goodies," she said, attempting a sprightly tone, unable to forget Jackson's wrath.

Jackson regarded her a trifle too long before saying, "Only the best for our aquanauts." He walked to the front of the speaker's table. "Please help yourselves, then take a seat." The members of Command Center took seats behind the tables. Once everyone was seated, Jackson began his welcome. "I'd like to introduce three members of our ROE team." He cited their countries of origin and scientific specialties before asking for applause for their service. To his credit, he kept his welcome speech brief before asking for comments on life on Yellow Submarine. In particular, he wanted to know their thoughts on whether humans could live their entire lives on submarines. When no one spoke up, he asked, taking Elizabeth by surprise, "Elizabeth, your thoughts?"

"Consider that I grew up on a farm where I had room to roam, so I did feel cramped at first, but not unbearably so. I imagine that a baby born there would never suffer as she wouldn't know anything different." She reflected for a moment. "That said, I advise that future designs be more spacious."

"But not by a large amount?"

"No, a large amount is probably unnecessary."

"Part of the issue is privacy," Pierre spoke up. "The public areas are large enough to accommodate all of us, but there are few places we can go and read a book in private, even in the library. Our bedrooms, if one can call them that, are confining."

"We could use a larger kitchen," Hans said, then added, "and more freezer space in case we happen to trap a whale." Command Center members looked confused but then chuckled.

Elizabeth refrained from shaking her head. Hans will be Hans no matter where he is. She felt the need to add something positive. "The food is good considering the limitations. Hans manages to cook chicken to make it taste different every time."

Hans stood up and took bows in several directions. Elizabeth resisted the urge to grab him by the ears and upbraid his lack of attention to the situation but said,

"That's our Hans. He has a monopoly on humor on board. I appreciate having instant access to doctors, dentists, libraries, and laundry rooms, but best of all is that circular walkway with a view of ocean fish. A walk there soothes my soul."

Pierre evidently had the sense to remain silent in the wake of Hans's weirdness.

Director Jackson ended the discussion by thanking the team members for their input. Elizabeth noted his satisfaction seemed genuine, and remembered the old bromide that honey attracted more bees than vinegar.

"I'll be here for another half hour if you'd like to express private concerns," Jackson said. 'There's a box by the door for your questionnaires and individual packets with your airline tickets and flight times for those not staying in New York. Have a good break."

Elizabeth swallowed her pride and approached her nemesis. "Director Jackson, I want to apologize for my behavior on the scaphe and thank you for abandoning the idea of bringing Alon to the surface. Alon is the name the team chose for him, and I feel we are friends."

"Well, thank you." His face beamed pure pleasure — not getting that her praise was backhanded. "And when I discovered the truth of the alien's ability firsthand, I forgot about the destruction deep sea specimens undergo when brought to the surface."

Elizabeth did her best to hide the repulsion that flooded her mind over comparing Alon to a specimen. She heard little of his launch into praise for her discovery and its importance to humankind yada yada — talking longer than in his welcome speech.

Despite the obvious implication, she ventured, "You must realize that it will be difficult to design medical equipment to study living things that obtain their oxygen from using gills in seawater?" A blank stare told her all she needed to know. Her feelings had alternated between hot and cold at this event. Now, she felt sick.

"I don't think that will be. . ." he mumbled, then said, "I need to answer other people's questions." He turned to Pierre, waiting behind Elizabeth, who must have heard every word.

"Has any progress been made on the vehicle's design to study the aliens?" Pierre asked.

"All I can say is that engineers will manage all that."

Elizabeth overheard and mentally added 'i.e., no progress, but we're still going forward by gosh and by golly.'

Chapter 43
(Debriefing with Pierre and Hans)

After the meeting, Elizabeth, Pierre, and Hans ate an early dinner at their hotel. Each had communicated with at least one of the members. "I spoke with three members at once, and none of them were excited about the vague plan," Hans said. The tall woman with long dark hair intimated that she suspects the director wants some credit for the discovery," Hans continued with incredulity beaming from his face.

"One member predicts it will never happen," Pierre said, to Elizabeth's delight. "There's not much we can do about Jackson. Let's enjoy our dinners,"

"Yeah, no reconstituted mystery meals for a while," Hans said. Pierre and Elizabeth laughed at Hans putting himself down.

Elizabeth forced herself to forget Alon for the moment and enjoy the evening. Hans had an early morning flight to Germany and left the restaurant. Pierre and Elizabeth retired to the dark, wood-paneled hotel bar. They found a dimly lit corner table away from the noise of those perched on stools watching the New York Mets play the Cleveland Indians. An old waiter with a shiny bald head and a collar of frizz approached them. He said nothing but stood expectantly with a pad and pencil. Elizabeth ordered a decaf coffee with a splash of Baileys, and Pierre ordered a Cointreau.

"How are you feeling?" Pierre asked after the waiter shuffled off at a snail's pace.

"Numb," she said but shook her head as if amused. "Maybe I shouldn't worry about Alon so much, but it's been my experience that what I worry about and work to prevent turns out to be a waste of time and energy, as it wouldn't have happened no matter what I did. Then, when I think that something bad has at most a remote chance of coming to fruition, it blows up in my face."

Pierre chuckled. "Then I suppose you have no choice." He placed his hand over hers. "Your report on an amazing scientific discovery is well written. Not only will American news want it, but all of the world. You may become a TV star." He made a sweep of his arm.

"Oh, please. But thanks."

"Yes, Dr. Pierre prescribes a compromise: do what makes sense, but then don't worry."

"Thanks, Doctor Pierre, I'll take that prescription, but doctors need to follow their own prescriptions."

Pierre chuckled. "Touche'."

"You seem to be less of a worry wort over the fungus recently." She found him easy to be with again.

Pierre ordered another Cointreau. Elizabeth hadn't finished her drink. She smiled as the walking-dead waiter made his way to the bar, then felt guilty about her amusement. "It's a shame the man still has to work at his age."

"Au contraire. It's what keeps him alive," Pierre smiled knowingly.

"You're probably right, but I held my breath when he placed our drinks on the table. A fly in the drink would get seasick."

"So, tomorrow. Let's meet for breakfast, and you can point me to places to visit in the US."

Elizabeth nodded. "Uh oh, here he comes. Let's hope you get your full measure of Cointreau."

Pierre turned his head slightly, and both watched the waiter's feet slide along the floor. He looked like a penguin in his black pants and vest with a white ruffled shirt. Pierre turned his head back before the waiter made it halfway.

"Couldn't bear to watch?" Elizabeth whispered, containing a laugh.

Pierre put his finger to his lips, and Elizabeth sipped her laced coffee.

The Cointreau may have been sloshing as it landed, but not a drop was spilled. Both smiled.

"I love it when you smile," Pierre said, surprising Elizabeth.

"We've both been too serious lately." She paused. "That's what we should have put in our questionnaires. The station has too little to promote fun things to do. We need a better variety of movies. I wish they weren't so paranoid about security, and we could email our friends and relatives back home for non-emergencies."

"I suspect a big part of it is that people would spend too much time on their devices," Pierre said. "The fact it's for security reasons is probably an excuse."

"Oh, I hadn't thought of that." She took the last sip of her coffee and sighed. Sitting in a fine restaurant with someone whose company she enjoyed had been a treat.

Pierre finished his drink and waved for the bill. "Now I need to stay awake until Mr. Slow Poke comes. It's been a long day. I'm tired, but I will walk with you to your room. New York is a dangerous place, yes?"

"All American cities are dangerous, but this hotel is nice."

"Nevertheless, it is no trouble."

The waiter spoke for the first time, asking for their room numbers. They signed their tabs and strolled to the elevators over dark-red patterned carpet. They discovered they were on the same floor — seventh. "Meet downstairs at nine?" Elizabeth asked.

"Make it 9:30." He kissed each of her cheeks and walked down the hall.

She wasn't tired and planned to polish her Alon report but found herself unable to focus for some reason. She went to the window and stared out at a lemon slice of a moon and realized how much she missed living on Earth's crust. She refused to believe that Pierre's polite pecks were anything more than his culture's traditional behavior. He was right about one thing. It had been a long day, and her throat still bothered her. She should have worn a mask on that walk on the beach. She gargled and climbed into her luxuriously large bed, stretching out her arms and not hitting a wall.

Chapter 44

(Elizabeth polishes her Alon report)

Rain softly pelting against Elizabeth's windows woke her in the early morning. She got up, settled herself at the room's polished dark wood desk, and tackled the changes she wanted to make to her story of Alon's discovery. It dawned on her that she needed to stop thinking of it as a scientific report and write it from the point of view of having answered whether we were alone in the universe, that our fellow beings have been only miles away. Once again, the discovery happened while seeking answers to other scientific questions. She liked this approach as Congress often questioned the value of researching questions they deemed obscure. Further, her paper was the perfect spot to credit Director Jackson in a manner that could trap him by creating an expectation of a better means of researching Alon's kind than he planned. She used Alon's name and described how he reminded her of human scientists. She lamented her limited ability to share the nature of life on top of Earth. His yes-or-no questions could only go so far in his desire to learn about human life. She included details about Alon's fear that his repressive government might not react well to interacting with other beings. She wrote matter-of-factly that any attempts to interrelate would be difficult and likely to succeed only in a long time in the future. Her feelings that Alon should be treated with utmost respect must be abundantly apparent.

She reread the entire article, wondering if a reader would recognize the importance of researching means of communication with or rather to Alon and also recoiling at the notion of physical examinations of Alonites. She asked people to consider how they might feel to be put under a microscope and prodded by beings from another world. Over and over, she conveyed in different ways that we are the aliens in the view of Alonites. A glance at her watch startled her. She had twenty- five minutes to get ready to meet Pierre. Instead of rushing, she phoned his room. He sounded relieved to change the time to 10:00.

She hurried anyway and had time to stop in the hotel's computer room, where she printed her latest version from her tablet. Special tablets were provided to the team so they could print letters in Yellow Submarine's situation room, which were eventually delivered to friends and relatives. Her sisters didn't write often, but when they did, their letters involving their children's antics amused her. The news about her mother rarely changed, which she had to view as a good thing. Dementia doesn't improve over time.

When she arrived at the coffee shop, Pierre waved to her from a booth. She slid in across from him.

"You look energetic this morning," Pierre said. He already had his coffee and a glass of grapefruit juice.

"I'm feeling good about the changes to my article on Alon. I brought you a copy; the changes are underlined." She picked up her menu.

A pert waitress bustled to their table and took their order: chocolate croissant for Pierre and a croissant sandwiching a fried egg and ham for Elizabeth. "How's the grapefruit juice?" she asked Pierre.

"Fresh squeezed and very good."

"I'll have one and coffee." Her hunger abatement on its way to gratification, she laid the report on the table.

Her coffee came quickly from another server, and she sipped slowly, watching Pierre. His face revealed nothing. "This is excellent," he said when finished.

"Do you think it creates an impression of Alon that would disgust a reader to learn he would be medically examined?"

"Disgust might be too strong, but a reader would . . . how does one say . . . be turned off, maybe dismayed."

"I'll take it." Elizabeth smiled. She felt energized and looked forward to a day in *New Yawk* City.

They quietly enjoyed their breakfasts. "So, what would you like to do today?' Elizabeth asked. "We can see some of New York while discussing the rest of your vacation in America. I suggest the Guggenheim Museum of Art to begin with."

"Perfect. Meet in the lobby in twenty minutes?"

"That works for me."

When Elizabeth entered her room, she saw a red light flashing on her phone and picked it up to hear, "You have a message at the front desk." Something inside her froze, telling her it was not good news. She decided to wait to pick it up at the end of the day.

Chapter 45

(Elizabeth and Pierre tour NYC)

They took the subway to the closest stop to the Guggenheim. On the way, Pierre proposed, "Let's agree to talk about the threat to Alon only at breakfast and enjoy the rest of the day."

Elizabeth realized from the pleading look on his face that she needed to tone down her obsession. "Agreed, but maybe just a little at lunch." She held her forefinger a half inch from her thumb.

"Have it your way, so long as we have dinner carefree." He shook his finger as if he were a parent and she the child.

Elizabeth merely grinned, but she understood he was serious. She had toured the Guggenheim when she went through her training, but Pierre had trained in London. She wondered how much he knew about this fabulous museum. It would be a challenge for her to describe how wonderful it was. Further, she wanted him to experience it exactly as much as possible as she had — completely unaware of its structure. The surprise had made her appreciation greater. One thing she liked about New York City was that there were so many different crossing subway tracks. No matter where you wanted to go, there would be a stop nearby. Even their Manhattan hotel was a five-minute walk from one. With a bit of luck, they could eat lunch in Central Park. When she had been here for training, it had taken her time and several instances of getting off at the wrong station to finally master getting around. Everything seems easy once you know how. They had set out early, wanting to be first in line as many people reserved in advance. All of the special tours were filled up.

Pierre insisted Elizabeth sit in an open seat near an overhead strap. Elizabeth rolled her eyes. "You know we women can't have it both ways."

"What do you mean?" Pierre asked.

"We can't claim equal rights to jobs, education, offices, sport money, etc. and expect men to be gallant about standing so women can be more comfortable."

"So, you think I'm gallant?" He smiled.

"Yes."

"It is the French way. You think people on this subway would not raise eyebrows if you stood while I sat?"

"I don't know. We'd have to try it."

"I think you're just having fun with me."

She snapped her fingers near her head. "Darn, you got me."

They were quiet for the rest of the ride. It was too noisy to talk, and the subway became crowded as more people came on than left at each stop. The entire ride was a little over a half hour, and both sighed with relief to get off into open space. "I guess our sleeping quarters on Yellow Submarine aren't the only cramped places in the world," Elizabeth said. She had outlined the walking route on a map and led the way.

"I have heard good things about this art museum," Pierre said. "I understand the collection is regularly updated as most do, but that there's usually at least one Van Gogh on display," he said in a questioning tone.

"I think so. I love Van Gogh. I had a picture of The Starry Night in my dorm room when I was a senior. It's too bad Van Gogh wasn't appreciated during his lifetime."

"That's too often true. Do you have a favorite artist?"

"Monet," Elizabeth said without hesitation.

"Oh, you must visit Paris and go to Giverny. You will see the ponds that inspired his lilies and his house where he lived for around forty years. He died in that house. The grounds themselves form a tangled beautiful park."

"I didn't know that. I thought Paris contained most of his work."

"It does, but Giverny is where he did his painting, so you are not wrong."

When they arrived at the Guggenheim, Pierre stopped and stared. "Wait a moment. This is not the right place. This is tres moderne. I've seen pictures, and the museum building looks European. This building looks like a giant flowerpot.

"You're thinking of the Metropolitan Museum of Art. We can see that too if you like, but another day. We also need to see some of Central Park today since we're close." They weren't first in line, but only six others were waiting. Chances were good, even if they had to tour at a later time. They could go for a walk or a second cup of coffee. They hustled to beat the occupants of several yellow taxis pulling up. The ticket office opened ten minutes later, and they made it! There was a special exhibit by Winona Silverstone in the Atrium. She had painted glaciers from all over the world before they collapsed or melted. With all of the painted ones now extinct, so-to-speak, her paintings were especially prized. They were beautifully sad. Elizabeth thought the melting streams looked like tears and guessed it was deliberate. They stood for a few moments looking up before boarding the elevator to the top. Once at the top, they were to circle down hallways with artwork on the outer walls. They could look across to see an individual painting again as they circled. "I don't know if you'll agree," Elizabeth said, "but I think this wonderful building is as artful as any of the paintings here."

"How could anyone disagree?" Pierre asked. After a few spirals down, he added, "It is also most practical. One doesn't waste time backtracking or searching a map to see where to go next."

Their tickets were only good for three hours, but despite the time passing quickly, they'd seen many of the best artists in the world. Picasso's Woman with the Yellow Hair — more peaceful than most of his work, — a soft Pissarro, Van Gogh, Gauguin, and several up-and-coming artists. Of course, Andy Warhol was a US representative who, in Elizabeth's opinion, wasn't in the same league as the French artists. She thought paintings by JoAnne Carson could compete, but she had never heard of her before. Why not a Georgia O'Keefe?

They left the Guggenheim, smiling. "I usually find my legs get tired after an hour or two in a museum," Elizabeth said. "But here, you're not standing still so much because you can view while you are walking, and I'm fine.

"Moi aussi," Pierre said.

Elizabeth smiled. She had to pass exams in two languages for her PhD. French was one, and she enjoyed Pierre's infrequent bits of French. She didn't think you needed to know much French to interpret his responses.

"There's a famous restaurant in the park, probably a half hour's walk inside the park, but let's take the subway and walk for less than a half mile."

Elizabeth was not surprised that they couldn't be seated at the Tavern on the Green, but they could order sandwiches and drinks to go. They picnicked under a nearby tree. After that, they wandered past large ponds populated by egrets, black-crowned night herons, and several types of ducks. The pond's surroundings and the pond itself made one forget the damage climate change had wreaked upon Earth. They strolled leisurely past statues and homeless sleeping peacefully behind bushes. They returned to the hotel in time for a nap.

Chapter 46

(Elizabeth and Pierre at Dinner)

When they met for dinner, they sat at Elizabeth's favorite corner table. "It appears that people in the US love their pets as much as we French, "Pierre said. "I saw many dogs in the lobby.

"Right. We always had a dog when I was growing up, but they weren't viewed as part of the family. They had to carry their weight: keep foxes away from the chicken house, round up cattle, and be living doorbells.

"So, you must train the dog to do this?"

"Yes, and how easy depends on the breed. My father made the mistake of getting a Dalmatian once. Not only could he not herd, but he kept killing chickens for the fun of it. One recipe for a cure was to tie a dead chicken around the dog's neck. In theory, as it rotted, the smell would stop the dog from its murdering ways. It didn't work. Farmers can't afford such a dog. He drove it to a pound in Falls City. It was hard, but I learned not to get too attached to our dogs."

Pierre reached over and put his palm over her hand. "Your mother is still alive?"

Elizabeth appreciated Pierre's attempt to change the subject, although her mother was not a pleasant topic either. "Yes, but as she once said, 'I am not with myself anymore.' She's in a memory care home. Whenever I see her, she tells the same story about when I was about to start school, and she planned to stand with me at the end of the driveway to wait for the bus on the first day, but my dad wouldn't let her because I needed to learn how to be independent. I sabotaged her argument by agreeing with him. I don't think she ever forgave me."

Pierre smiled, "So, you were independent at an early age."

"I guess so. I remember waiting for the bus every morning. I would watch the horizon trying to determine the exact second I could see a strip of the yellow- orange bus."

"Ah alors, the curious scientist at an early age?"

"Huh? That's a leap. What about your parents? I know they're still alive."

"Since I was the youngest child, once I left the nest, my parents were freer to become volunteers for the welfare of the world. They'd always volunteered but only a couple of weeks at a time. Now, their trips are longer. Currently, they are in South Africa volunteering to help underprivileged children in the townships of Cape Town and Johannesburg. Even if I took time to fly down there, they'd have little time to spend with me. That's why I'm using my break to see more of the US."

When their meals of duck a' la orange for Elizabeth and beef bourguignon for Pierre arrived, their conversation abated, although Pierre mischievously joked about the duck coming from a pond at Central Park.

If Elizabeth could have reached him, she would have punched his arm.

They shared chocolate mousse for dessert not wanting the evening to end. Once the waiter had them sign the bills to their rooms, they felt they had no choice but to leave. Pierre took Elizabeth's hand as they took the elevator to their floor. This time he lingered during his pecks on her cheeks. "Thank you for arranging a wonderful day," he said, then kissed her full on the lips with an embrace that she also never wanted to end. They stared into each other's eyes before Pierre left for his room. Finally, he said, "Breakfast tomorrow at 8:00?" He blew her another kiss.

Elizabeth had to remind herself of Command Center's Rule #9.

Chapter 47

(Elizabeth receives bad news)

Elizabeth woke up early and remembered she'd never checked her message from the front desk. She discovered her sister, Suzanne, had called. She feared the worst and wasted no time returning to her room's privacy to talk. "Hi Suzanne, I'm sorry I didn't get back to you sooner. She started to explain, but Suzanne broke in. "Mom had a severe stroke. She's not likely to live many more days."

Elizabeth slumped and couldn't speak for a moment. "OK, I'll come home as soon as I can."

"I've already checked flights for today. Any chance you can get on one to Falls City on United five hours from now? Otherwise, eight hours into Omaha?"

Elizabeth thought for a moment. "There's no reason I can't take the one in five hours, and Falls City is closer to home."

"Great, I reserved both. Give me your credit card number, and I'll handle it."

Suzanne was efficient, as were both her busy sisters and dedicated mothers. After Elizabeth gave her the information, Suzanne asked, "I assume Mom is in the Falls City General Hospital. Is she conscious?"

"Just barely, more like living in a daze. She hasn't a clue about what's going on. When I tell her I'm Suzanne, she doesn't respond nor seem to know who that is. Marjorie flew in last night. She's staying at Johnny's. She'll ride up from Dry Creek with him later. We're taking turns being with her. Johnny is the one who found her in trouble in her room in Remember Always. I'm staying at a hotel within a ten-minute walk of the hospital. There are two single beds, so you can stay with me unless you want more space."

"Are you kidding? That will be luxurious compared to my space on Yellow Submarine. How is everybody holding up?"

"OK," Suzanne said with a choked voice. "We all knew this would happen someday. I'm surprised she lived as long as she has. How many live past ninety-four?"

"Right. Well, I'd better get going and let people know I have to change our plans."

She was late meeting Pierre.

"Oversleep? he joked until he looked at her face. "Something wrong?"

"Yes, my mom is in the hospital and not likely to live long. I'm leaving this afternoon. I'm sorry I won't be able to tour with you."

"Don't worry about me. I'm sorry about your mother. How old is she?"

"Ninety-four. She's been declining for the last four years." Elizabeth stiffened her lips. "I think I'll just have coffee."

"No, you should eat something substantial. You have a long day ahead of you, and you don't want to rely on airplane food." He reached across the table and put his hands over hers.

"You're right. Have you ordered?"

"No, but I want to try their croissant egg sandwiches."

The chirpy waitress had arrived and overheard. Elizabeth handed over her menu and said, "Make that two."

Elizabeth insisted on giving Pierre as much information as possible about the places they had planned to visit, including leaving her maps and brochures with him.

If only her mother could hold on so she can see her once more.

Chapter 48

(Elizabeth travels to her roots)

She arrived at the airport several hours early and purchased a temporary phone for a month. Cell phones weren't allowed on Yellow Submarine, supposedly for security reasons. She added her siblings' phone numbers from her old book of addresses. She enjoyed a bit of window shopping and walked the corridors, knowing she'd be sitting for hours.

The flight to Minneapolis was on time and uneventful. She had an hour before boarding her next flight. She found something cheerful for her mother in one of the little shops that lined the path to her gate. She couldn't decide between a snow globe with a reindeer and a green porcelain frog, so she bought both. If her mom didn't appreciate them, one of her nephews might like them. The flight to Falls City was a small plane with two seats to the sides of a single aisle. Her seat was adjacent to the window and beside a rotund man in a too-tight business suit. She was glad the flight was only an hour as she sensed her seatmate might be a pain. She pulled the plane's magazine from the pocket in the seat in front of her and located a crossword puzzle.

"I never look at those magazines," her seatmate pronounced, and she groaned internally. "The stuff they have is way too high priced, especially when I know where I can buy everything at a bargain, and if it's not a bargain it will be by the time I deal with the salesperson. Want to know my secret?"

"Not really. If I want something, I only worry about getting good value for my money."

"It's always a good value if you can get it as cheap as possible."

"Not if it wastes your time. I'm sorry, but my mother is on her deathbed, and I'm not feeling up to a conversation right now."

"Oh, I know how you feel. My mother's death was hard on me. I was the youngest of four children, and I was Mom's favorite. She always …"

"Excuse me, I need to use the bathroom. I'm feeling sick."

"You can't on these short flights."

Elizabeth called for an airsick bag.

"I'll also get you some water," the flight attendant said. "Is there anything else I can get?"

"No, I think I'm OK now. I just need to stay quiet."

At first, she thought her ruse had worked, but five minutes later, the oaf began again.

"Feeling better now?" he asked but didn't wait for an answer. "What makes me sick is the damned government. Them Democrats want to spend like there's no tomorrow. Always wanting to try some crazy new thing to fix the environment."

"Yeah," Elizabeth said, "Who wants to breathe clean air and have clean water to drink? Then there are those extreme temperatures. Who cares about that anyway?"

Rotund guy's mouth dropped open, and his eyebrows furrowed. He sat quietly for an entire minute. "You must be one of those liberals."

"Guiltier than sin. One definition of a liberal is someone who lives and lets live, so I'm going to let you live, but if you don't mind, I'd like to spend the rest of this flight in peace."

The rotund man huffed. "OK, then. Geez, I was just trying to be friendly."

Elizabeth turned a page in her magazine, sighed, and relaxed for the rest of the flight. Suzanne met her at baggage claim, and asked, "How was your plane trip?"

"Good, for the most part. Everything was on time, but I sat next to a major bore from Minneapolis to Falls City."

Suzanne chuckled, "I hope you got him to leave you alone."

"I finally did," she shook her head. The woman who sat in front of him gave me a thumbs-up as we were walking to baggage claim."

"Do you want to go to the hotel and freshen up or go straight to the hospital?"

"Straight to the hospital."

She wondered whether she'd recognize her mother as the mom she grew up with.

Chapter 49

(Elizabeth's mother's time is limited)

"I hope you're prepared," Suzanne said as they took the elevator to the third floor. "You haven't seen her in nearly two years."

Elizabeth said nothing and just followed Suzanne to the room. Her mom had aged significantly with pockets below her eyes and sunken cheeks. Her eyes were open, but it wasn't clear she was fully awake. Then she began to blink. "EliZAbeth," she exclaimed in a voice she had fun using when Elizabeth was a child. Her mom didn't think of her as an adult because she wasn't married. Suzanne's jaw dropped. Then their mother began to spew, "Graduate school, big protester, in jail, Suzanne's boys, Margaret's husband." She squirmed to the side of the bed. "Going home."

"No, Mom, you can't."

She repeated more emphatically "Going home."

Suzanne told Elizabeth to get help while she tried to keep their mom from falling out of the high hospital bed. Elizabeth ran back to the nurse's station only to find no one there, and a sign "Back in five minutes." She scanned the hallways and saw a nurse down one of them. She practically accosted her, "My mom is in 329 and is trying to get out of bed to go home."

"I'll send someone right away." She picked up the autocall around her neck, and spoke into it, "Alert, someone needs to get to Room 329 ASAP to contain the stroke patient." She turned to Elizabeth. "Sorry, I can't help right now, but let me assure you someone will be there promptly."

In fact, by the time Elizabeth got back to the room, she heard footsteps behind her. Suzanne was hanging onto her mother's flailing arms. Two male nurses ran to her side. One took her left hand and picked it up. "Pretty ring, Violet," he said, and her face brightened, as she nodded and held up her hand. The other said, "You look uncomfortable, let me help you sit back up." The two easily repositioned her, then stood by. The first aide said "Your ring looks like a classic wedding ring from decades ago. You are lucky to still have it." Violet held her hand up again and looked at her ring. "Pretty ring."

The second asked, "OK if we give her something to calm her down?"

The sisters looked at each other. "What something?" Elizabeth asked.

"Temazepam, it's a mild antidepressant, which relaxes you and can help you sleep. It's a pill, which we can mash into her water."

Elizabeth looked at Suzanne who nodded. "We're OK with it."

The second nurse left saying he'd be right back. The first waited to ensure the patient wouldn't try to leave again.

Soon, the drama was over, and the sisters went to the coffee shop, while their mother slept.

"Now, I know what you've been dealing with, "Elizabeth said.

"This was the worst. On the other hand, she was talking, and it made some sense. You know that she never understood why you went to graduate school."

"Yeah, I guess so."

"You probably don't know that Mom and Dad argued about it. Dad thought you shouldn't get married because you were so smart, but Mom didn't agree. Dad didn't understand that one could do both. They were both embarrassed when you were arrested in D.C. You shouldn't have told them that the police bloodied your nose and neck because then they knew you resisted. When the story came out in the Dry Creek Bee, the whole town knew it." She paused. "Mom and Dad both wanted to keep up a good face but in separate ways. Later they were proud, or at least Dad was, that you were a leader in the movement. That's once it got going, and the youth forced the government to act. Then he claimed you took after him. Remember he was once kicked out of the back room of Rusty's Dusty Bar for arguing politics?"

"Yeah! How do you think they were different about maintaining a good face?" Elizabeth asked.

"Dad would never miss a funeral or a church gathering believing attending was an obligation. What would people think if they didn't go? That's how 'What would people think' became our family motto. On the other hand, he'd make fun of Mom in front of other people, which humiliated her. Of course, Dad was just teasing. Mom was jealous of her sisters whose husbands bought them expensive jewelry, and romantic cards. Dad would tell people he needed to check with the 'old lady' as if it was just another term for 'wife.' He wasn't the only man who did that, but ..." Suzanne tipped her head. "Do you remember Dad always teasing mom about Alfonso? He had to be the oldest guy at church.

I thought it was because the guy was the least likely man to attract Mom, but later, she told me that as a young wife, one day, she was hanging some rugs on the fence, and Alfonso stopped by to see Dad, who wasn't home. He told Mom that he couldn't sleep at night thinking about her. When she told Dad, he teased her for the rest of her life about him: He thought it was soooo funny."

Suzanne began laughing through her tears.

"I'm sorry," Elizabeth said.

"No, don't be. It's so good to see you and reminisce about the good, the bad, and the ugly."

"Do you know what time Margaret and Johnny are coming?"

"Any time. They could be up in her room now."

"Let's go up then. I want to catch the doctor. I wonder if that episode means she's not as bad as they thought. In fact, you never gave me any details beyond the stroke."

"Ask the doctor when we see her. Massive was all I heard, with an extremely poor prognosis. That's not to say I understood every medical term."

They encountered Margaret and Johnny heading for the elevator. "Fancy meeting you here," Johnny said. Elizabeth believed he'd inherited corniness from their dad. Margaret gave Elizabeth a hug, her face saying all that needed to be said. Has anything changed since yesterday?" she asked.

Suzanne chimed in. "You won't believe what happened. She related the story of Mom saying EliZAbeth exactly as she used to do.

"Oh, yeah, I heard enough JohNAthons in my lifetime."

"I don't remember any MarGArets. I guess I was the good kid." Margaret preened in fun.

Their mom was still asleep when they got to the room. "How's the coffee down there? I'm tempted to get a cup." Johnny asked.

Suzanne said it was not bad, and Johnny asked if Margaret wanted a cup, but she shook her head. The three sisters stood silently by their mother's bed.

Elizabeth picked up a small book of photographs on her bedside stand. They were all of her grandchildren. Elizabeth came upon her favorite picture of Bob. "He looks like Little Lord Fauntleroy in his outfit and arm resting royally on the armchair."

"Looks are deceiving," Suzanne said. "He was such a terror at that age, always in motion. I think the photographer must have velcroed his arm to the chair."

Margaret sighed, breaking from her reverie. "I knew this day would come, but that doesn't mean I was ready for it. Johnny seems ready. He told me her Alzheimer's had been getting worse. He stopped in to see her about once a week. He's as close to Remember Always in Sioux Ville as he is to Dry Creek. He and Callie buy groceries there, so I guess it's convenient to stop in, but there hasn't been any real communication in some time.

Elizabeth found herself feeling less guilty for being in absentia for as long as she was. A nurse came by and said, "Only two visitors are allowed in a room at a time."

"We're her daughters, we've come from some distance, and she's asleep," Suzanne said.

Johnny appeared in the doorway, coffee in hand. "And I'm her son,"

"All right, just keep her calm if she wakes up. Dr Spier is on the floor. She'll be here soon."

Dr Spier hustled in a few minutes later. "Is this the whole family?"

"It's all of her children," Johnny said.

"Let's talk in the lounge. No one is in there," Dr. Spier waved her hand in the right direction and followed them there.

While getting seated, Elizabeth asked, "Were you told she had a moment of near lucidity and fought to get out of bed to go home?"

"Yes, I saw that on her chart."

"Is that a positive sign?" Elizabeth asked.

"It may seem such, but it's not that uncommon for this to happen especially if there's some kind of trigger. I understand she hasn't seen one of you in some time."

"Can you tell us all in plain English about her stroke besides that it was bad?" Johnny asked.

"I'll try," she said smiling. "I told you, Johnathon, that she had a hemorrhagic stroke. That means a blood vessel has burst in her brain. Your mother's was in a critical place. Given her age and history of Alzheimer's, her chance of recovery is next to nil. I expect that her entire brain will shut down, including the parts that govern breathing, heartbeats, kidneys, liver, and lungs, that is, her whole body. I understand she has a DNR."

No one spoke, but they all nodded.

"I suppose it's unfair to ask how much time she has," Johnny asked.

"No, but you can't hold me to it. The human body continues to surprise me. The glib answer is 'not long' but it could be days, or it could be weeks. It's possible but highly unlikely to be longer than that."

"Is she in pain, do you know?" Elizabeth asked.

"Probably not physically, but mentally she is likely suffering confusion and disappointment that things are not working the way she wants them to."

When everyone quieted, Dr. Spier said "I hope I've answered your questions. She can die peacefully knowing she has a lovely family."

Tears trickled down three sets of cheeks. Johnny turned his head to the side. They sat in silence, trying to come to grips with their mother's imminent demise.

Elizabeth hated herself for thinking about returning to New York and coming back for the inevitable funeral. She also regretted thinking for the last few years that her mother was already gone.

"What should we do now?" Margaret asked.

"Go have pizza and talk about it," Johnny said.

The girls, as Johnny referred to his sisters, agreed, and they found out from Johnny that their mom had written down years ago what she wanted at her service down to the last pallbearer and the menu for the food following. Both Suzanne and Margaret expressed the need to return to their homes after a couple of days and return later relieving Elizabeth of some guilt of how long she had been away. If only she could be certain that her visits would not have mattered?

Margaret and Johnny left for Dry Creek, and Suzanne and Elizabeth returned to the hotel, each in a semi-stupor and steeped in their own thoughts.

Chapter 50

(Violet Gayer is no longer suffering from Alzheimer's)

When Elizabeth and Suzanne returned to the hospital the next morning, their mother's bed was empty. A nurse passing in the hallway stepped in. "Your mother has gone to a better place." This was a line Elizabeth hated even when she was twelve years old. She wondered if the afterlife was a better place, why no one wanted to go there? Now, however, the consequences of the phrase hit her in the belly. She and Suzanne instinctively hugged each other.

"We'd better call Johnny," Suzanne said. "He'll notify Moeller mortuary."

A hospital clerk approached them gingerly. "I'm sorry, but there's paperwork to be signed. Do you want to see your mom first?"

Elizabeth thought for a moment, then said, "I think I'll pass."

"Then I will, too," Suzanne said.

They took care of the business, made the phone calls, returned to the hotel, packed up, and drove to Johnny's — the farmhouse where they were raised.

Johnny's wife, Callie, greeted them at the door. "It's happened sooner than we thought," she said, shaking her head. "Come on in." She guided them to the kitchen table.

The sisters took a deep breath, and Callie nodded. "I know." After a moment, Callie began. "Johnny called Moeller's, and they'll pick her up later this afternoon. Margaret is on her way back from Omaha. She said she'd packed in such a hurry she wanted to go home and prepare for whatever came next, Johnny wants to finish his workday and talk to the guys who will be filling in for him. Our boys are in school. I won't tell them until they get home. Have you had anything to eat? There are cinnamon rolls left." Not waiting for an answer, she put a plate of them on the table along with the requisite coffee pot.

The sisters sat sipping coffee and nibbling on the rolls while Callie took care of calling the minister and a few others.

"Now we can remember Mom as she was," Suzanne said. They began with fond memories, even ones that were unpleasant at the time and those that described strong feelings. They remembered how generous Mom was with Christmas gifts and how important it was to be *fair*. If one of us had four packages to open and the siblings had five, she had to point out that the cost of the four \was as much as the other's five, as if anyone noticed in the room swimming with unwrapped Christmas paper.

"I don't know why our dad never bought Mom gifts," Elizabeth said. "It's as if he thought it was unmanly to treat his wife to something special or do anything romantic."

"It's related to how he was raised," Suzanne said. "Our Grandpa Gayer was quite something. Elizabeth, I'll never forget when you bought this romantic Christmas card and made Dad sign it."

"Oh, sweet deity, yes!" Elizabeth said. "He didn't even look at it. Then he signed it 'Harry Gayer!' I thought it gave away what I thought would be a joke, and we could tease Dad about how romantic he was."

"Boy, did that backfire," Suzanne said. As a joke, I mean. Mom took it seriously. Grandma Vander Stout was here, and she had tears in her eyes when she read the card. The fact that no man would sign a card to his wife, including his last name, escaped both of them."

"Yes, it ended up differently than I predicted, better in fact," Elizabeth said. "I expected Dad to deny he bought the card, and I would lie and act innocent. But he didn't. He just grinned and seemed to enjoy Mom's pleasure. Now, I wonder if he didn't understand the importance of small expressions of love, especially with the father who raised him. He did truly love her."

"I don't think men talked with each other about the gifts they buy their wives. It was a different era and a different culture." Suzanne said, picking up a cinnamon roll. "These cinnamon rolls *are* good. I wonder if Callie got them at the Honey Bun."

Callie heard her and sat down. "Yes, everyone knows they're the best. I get most of our groceries in Sioux Ville, but I have to make a special trip to Dry Creek to get these, and their hamburger buns," she added.

"Speaking of which," Suzanne said. "Don't plan any meals. We'll eat out or get takeout. Elizabeth and I have reservations at the GymJam Inn, so we'll be in Dry Creek if you want us to pick up anything.

"That sounds good, but we must use our house as a meeting place. If the boys don't get at them first, we have plenty of coffee, soft drinks, and snacks. You'll be surprised at how much Billie and Kenny have grown."

Margaret arrived, and the three sisters hugged. Fortunately, I hadn't started repacking. I just added clothes for the services and clean underwear," Margaret said. "I guess the timing is for the best with you here, Elizabeth."

Elizabeth sadly nodded.

When Johnny arrived home from work early, they turned to talking about growing up in this house and how they valued their upbringing.

"When I was working on my PhD thesis and was one of a few women in my classes, it occurred to me that I did traditional male chores ever since I turned eight, as well as the work females usually did. I think it gave me the sense that I could do anything I wanted," Elizabeth said.

"We all did," said Margaret, but yes, as the oldest, I think you did more outside."

"Well, I'm glad I had big sisters, so I didn't have to do any woman chores," Johnny roughly cracked with a smile, sounding exactly like his dad.

"Yeah, you always wanted to farm. It's funny that Dad thought he had to have a farm for each of us when, at the same time, he also expected us to go to college," Elizabeth said. We didn't do that to become farmers' wives."

Callie spoke up in defense of her role. "I love being a farmer's wife. I like being busy."

"I'm sure it does keep you busy, but it's a little easier now. You have a riding tractor to mow the houseyard grass and the orchard." Margaret said.

Callie agreed and expressed appreciation for Johnny bringing in gravel to cover the driveway and a path around the houseyard. They were tired of dealing with the mud in the spring and fall.

"The worst thing about being a farmer for me is that you can't plan economically," Elizabeth said. "If you have a good year, you'd have to wonder if you could spend money on a new car because next year's crops could be hailed out or get no rain."

"That's why Johnny has two jobs," Callie said, "and I work part-time at the Sioux Ville bank. "We can always count on that income."

Kenny and Billie burst through the front door, full of energy. They were excited about Halloween coming up and described their costumes: a clown for Kenny and a devil for Billie.

Callie calmed them down. "Do you know why all your aunties are here?"

Kenny, the older of the two, asked solemnly, "Did Granma die?"

Callie nodded.

"She was old, right?" Billie asked.

"Yes, boys, she was very old."

"Very, very old," Kenny said.

"I'm not going to get old," Billie said.

Kenny snorted. "Then you have to die young."

Billie began to cry, and Callie called to him. "Let's go upstairs and look at pictures of Grandma holding you when you were a baby. You too, Kenny."

"Hey, do you feel like driving into town?" Elizabeth said to her sisters. "I'd like to wander around the town square and poke in the old stores."

"I'd love to go," Suzanne said.

"I'm going too," Margaret chimed in. "Callie, is there anything we can pick up?"

"Just more cinnamon rolls."

Chapter 51

(Dry Creek)

The town square had been renamed Swanstrom Square for the detective who dealt with the legendary murders of Dry Creek. A plaque describing that horrific time had been mounted on a stone statue in the shape of a tall rook from a chess game. Elizabeth thought not enough credit in the town's history had been given to her ancestor, Soledad. Otherwise, nothing had changed. The bandstand was still there and evidently still in use for music and other public events. They strolled through Audrey's Apparel, Swiesau's Pharmacy — which still had bar stools where you could order milkshakes and find greeting cards on a spinning tower of racks — Van Sloten hardware with kitchen products as well as typical hardware tools, Notions n' Things — that always had unique items — and they peeked down at Tony's Barbershop, which one had to access via steps as it was under Notions n' Things. Dry Creek was a town that time forgot.

Elizabeth bought a simple peach-colored sheath at Audrey's apparel for the funeral and a beige skirt with abstract white and rust-red swirls resembling flowers with a white pocketed cotton blouse for the viewing. It seemed strange to shop for clothes, and she wondered if she'd feel right ever wearing them again.

The bowling alley had been usurped to expand the adjacent restaurant, Chuck's Chuck Wagon, the only restaurant in the town proper. Margaret offered to treat the family to dinner and dropped in to make reservations. A golf course had been added just west of town, which had a large restaurant that could be rented for parties and class reunions. The old hamburger place had been replaced by a second-hand store, Victoria's Treasures. It was one of the more interesting shops they visited. It shouldn't have surprised Elizabeth that older clerks asked if they weren't the Gayer girls since Dry Creek's grapevine must have informed everyone in town of their mother's death by now. They passed a flower shop that hadn't been there when they were growing up. Callie had told Margaret about it, commenting that the owner, Sandy **Schmidt,** would be doing a good business given their mother's funeral. In fact, funerals sustained their business. The town had a new beauty shop, Dress Your Tresses. Their mother used to have her hair set once a week in the same location. She wore hilarious fluffy hair nets to bed, reminding Elizabeth of Mrs. Piggle Wiggle. Once, when Elizabeth came home from graduate school at six p.m., her mother surprised her by wearing a nightgown capped by a fluffy hairnet. She had expected dinner, not that she had any trouble finding provisions for a sandwich. Later, she realized that must have been the beginning of her confusion about time. Before she went into memory care at Remember Always, she would head for church at midnight and wonder where everyone was.

On their way out of town, they stopped at Dry Creek's cemetery and visited their dad's grave. They were surprised to see that their mom's grave had already been dug in front of the stone next to their dad's. It quieted the sisters, and they left with tear streaks marring their makeup.

They stopped at Soledad's grave and walked past the nearby R I P metal signs. They shook their heads at the raven perched on the P. What if her ancestor had never existed? Would Dry Creek have continued to experience unexplained deaths?

Chapter 52

(Elizabeth and siblings arrange the services)

Time passed quickly. They chose a casket the rose color that their Mom had prescribed, held a viewing the night before the funeral, and chose the menu for after the service. They went with their mother's menu of ham buns and chocolate cake with chocolate frosting. The casket was placed at the head of a large room for the viewing. Everyone commented on how good she looked. When Elizabeth was young, she never understood how important that was; she just thought that if the body didn't look good, people would whisper to each other about what a shame that was. People sat in rows of chairs while the funeral director presided over volunteers offering their memories of Violet Gayer. Elizabeth did not plan to say anything, although she and the rest of the family always threatened to tell the chicken story. Her mother's church friends talked about her pride in how well she dressed her young daughters, often in matching dresses, and praised her dedication to the church. Elizabeth's sisters showed their appreciation for how she helped out after the birth of each grandchild. The somber mood in the room motivated Elizabeth to go through with her family's common threat. Everyone knew how gentle her mother was and how rarely she spoke in anger or even frustration.

Elizabeth stood and said, "I'm going to tell a story we always threatened our mom we would tell." For some reason, that brought laughter. "You all know my sweet mom. Despite the fence around our houseyard, the chickens would get in and scratch in the dirt and uproot Mom's flowers. Nothing infuriated her more than having to replant them. She never saw them at it, but one day, she was sweeping the front steps and spotted a hen in her flowers. She flew down the sidewalk and began to beat the hen with her broom. Round and round, she chased that cursed chicken until it died of exhaustion. My grandfather said he'd never eaten a more tender chicken." One of Elizabeth's uncles, Douglas, her mother's brother, laughed heartier than the rest. Elizabeth's heart warmed while her eyes teared.

The night of the viewing had been a welcome catharsis for Elizabeth, but the service made Elizabeth cringe. The pastor gave a long, tedious sermon. Never pass up a chance to convert the heathens, she imagined as his philosophy. The woman who sang didn't have the right voice for Amazing Grace, providing another cringe moment.

At the lunch in the church basement, she enjoyed catching up with the lives of numerous relatives. When someone tapped her on the shoulder, she expected another cousin but stared at her face for a few seconds before her jaw dropped. "Karen! Thank you for coming!" Elizabeth had forgotten that Karen had married the owner of the drugstore in Dry Creek. It was the first time she'd seen her high school friend since graduation.

"Well, you never come to our reunions, so I guess it takes a death in the family for you to come back to Dry Creek," Karen said. Your brother keeps me informed. I can't believe you're living on a submarine.

After a chat about old times, including Karen's updates on their old gang's friends: Maureen, Sharon, Helen, Janet, Dianna, and Sue. She archly pointed out that *they* actually came to class reunions. "I'll let you go for now. You must have lots of people to catch up with, but stay in touch."

Chapter 53

(Thank you notes and memories)

Johnny insisted they go straight to his house and write thank-yous for all the flowers and gifts. He wasn't about to get stuck with the chore. Elizabeth would have preferred to relax, but Suzanne took the list to organize and divide up the work. "Say, anyone know a Pierre? There's no last name or address. We know that he ordered through the Moeller Funeral Home's website."

"What? let me see if you have the name right." Elizabeth said.

"So, you know him, Johnny asked.

"Yes, he's a team member on Yellow Submarine. He's also on break. I was about to go with him to see a museum when I got the call from Suzanne. I can't believe he did this. Which flowers were his?"

"The white lilies," Elizabeth's siblings said at once.

Margaret cocked her head. "Anything going on between you two?"

"No, we work together and are good friends."

"Okay, we have a lot of thank-you cards to get through, so let's not get distracted," Suzanne said.

Sitting at Johnny's kitchen table, they shared information about the relatives and old friends they spoke with, helping them make quick work of the task.

"Did you talk to Steve Williams?" Margaret asked.

"Oh, is that who that was?" Elizabeth asked. She began to laugh. "I loved his mother, Nellie."

Nellie Williams had a figure like many women of a previous generation who had given birth to six or more children: legs and arms of normal size but a stomach as large and round as a beach ball with a bosom like two water balloons resting on top. Her full face always wore a contagious smile.

Elizabeth loved it when her mom stopped at Nellie's. Nellie would make you feel you had made her day. Farmers around Dry Creek once thought nothing of popping in at their friends' or neighbors' houses. Elizabeth laughed. "Remember that time when we dropped in, and she offered us kids candy bars?"

Margaret and Suzanne joined Elizabeth in laughter. "Yes, when she opened the cabinet, she groaned that Lyle, Tom, and Stan had nearly decimated her supply. She found only one candy bar, and it had a bite out of it. Nellie cut off the bite, cut the rest into three pieces, and gave them to us," Elizabeth continued. "Then she said, 'And the bite is for Mom' and handed it to her." I don't think we could say her name after that and not remember the candy bar incident."

"Yeah," Johnny said in the same gruff voice his dad often adopted. "I got tired of hearing that story."

Elizabeth's brothers-in-law had driven to Dry Creek with her one niece and five nephews, but they left immediately after the church basement lunch. Of course, Elizabeth mentally noted their growth but didn't want to chant the usual "My, how you've grown." Further, it was more maturity than growth in the case of Suzanne's sons. She liked both brothers-in-law. Her sisters had married well.

After the thank-you cards were written, addressed, and stamped, they returned to memories of their parents and how they both were political. Her dad was outspoken, while her mom wrote letters to the editor. Elizabeth mentioned her calling frequently and helping her with phrasing. "I wasn't as busy as you guys with your babies then. Mom took pride in having everyone comment on how good her letters were. Did she tell you she received hate mail over a letter where she commented on the sour grapes of the Republicans over Obama's second win? It was partly my fault because when she said, 'Those Republicans should just take their marbles and go home,' I told her, 'That's brilliant, use that in your letter.'"

"She wrote poems for every friend and relative who died," Johnny said. "That was after Dad died and before she went into Remember Always. She even wrote one for her 90th birthday. I still have a copy. Want me to get it?"

"Do cows shit in the barnyard?" Margaret asked.

"It didn't take you long to talk like a farmer again," Johnny said as he left to find the poem. "I hate that name. Remember Always," Suzanne said out of the blue. Her sisters agreed.

Johnny came back. "I found it. Want me to read it?"

"No, we were just testing whether you could find it," Margaret said with obvious sarcasm.

Listening to her mom's last poem was like listening to soft, sad music — a perfect ending to the family's reminiscing afternoon.

They relaxed for an hour before they needed to leave to keep their reservation at Chuck's Chuck Wagon. Johnny had access to their mother's checking account and pronounced that their meal was on Mom. It was a nice gesture, one Elizabeth knew would please her mother.

Afterward, with early flights the next morning, Elizabeth and Suzanne went straight to their hotel in Falls City. Suzanne lived in Minnetonka near Minneapolis. Margaret had driven from Papillion, a suburb of Omaha. The blend of pain, relief, memories, and connection left them all subdued and reflective.

Chapter 54

(Elizabeth returns to NYC)

On the leg to Minneapolis, Suzanne expressed her thought that things had gone as well as they possibly could, and Elizabeth agreed. "Funerals are not part of our daily lives in many ways, but they make us understand what's important. I'm so glad your boys could make it."

"Oh well, I thought they could have hung around longer. One ham bun and a slice of chocolate cake with a nod here and there, and out the door, they went. I suspect Richard put the fear of God in them to attend, but he probably had to promise to get them back to the airport in short order. Adam had midterms coming up, but the older three have jobs with good benefits. They could have stayed a couple of days."

"Well, on their behalf, I can't imagine they would have enjoyed helping write thank-you notes, Elizabeth said. "Do you think they would have enjoyed hearing our childhood stories?"

"Maybe." She paused. "Sometime, we should start our own reunion like those we went to as kids in a park in the middle of summer."

"And in the middle of fly season," Elizabeth added. When they both laughed, the man in front of them turned and gave them what, as kids they called the evil eye.

"They whispered. What did you think of Pearl's dress?" Suzanne barely mouthed. Elizabeth signaled with a thumbs-up. "Margaret has always said she was a fashionista."

"She and Mel seem close."

Margaret chuckled, "Most twins are."

"That was a duh. Where did they go after the funeral?"

"You were in the bathroom when they asked if they could poke around in town. I told them to go ahead. Margaret agreed but told them they had to be on time for dinner. They said they would never miss the steaks at Chuck's Chuck Wagon."

Time seemed to fly, and Elizabeth was surprised to hear the *landing preparation.*

Soon, she'd said her goodbye and sat looking out the window on her way back to New York. She pulled out the magazine she had taken from the trip to Minneapolis, turned to the crossword puzzle, and plugged in earphones playing soft music.

As soon as she landed, she regretted the time she missed with Pierre. By now, he'd be hiking in New England. She consoled herself with the thought that she needed to spend her time mounting a major campaign to protect Alon.

She was walking through Concourse B when an intercom called her name. "Elizabeth Gayer, please report to the United Airlines desk." She couldn't imagine why. Airlines never called you if they messed up your luggage. You had to track them down. She went directly to the check-in area and couldn't believe her eyes. Pierre was waiting nearby.

When they caught sight of each other, both faces lit up. "What are you doing here?"

"That's a rather *stupide* question," Pierre responded.

"But I thought you would have left for New England by now."

"I've already been there. I left the same day you did. Acacia National Park was amazing. I had to buy a warm jacket at the visitors' center for the hiking, but it was *merveilleux*. I found it calming to be outside and wandering through nature. Now I'm back to see more of New York." As if he realized how upbeat he sounded, he said, "I am sorry you had to experience dealing with the loss of your mother. That had to have been difficult."

"It's hard to describe. I lost my mother a long time ago. Now I can remember her as she was in her good years. That helps." Her words were positive but mixed with sniffles in the dough of her feelings. She gathered herself and continued. "It was good to reminisce with my family; we visited places I remember growing up. Dry Creek seems unchanged. Oh, thanks for the white lilies. That was a sweet surprise." She paused and took a deep breath. "I hope you didn't rearrange your trip for nothing. I need to spend time on launching a campaign to save Alon. I can spare half-days, though."

"Half-days sound good, and I can help you. At least I can be a sounding board."

Pierre had been much more than that, and she needed to suppress her feelings.

Chapter 55

(Elizabeth launches her campaign)

Elizabeth had been talking about climate change from high school through PhD thesis and beyond. Consequently, the words for her kick-off letter cannonballed from her brain to paper. She wanted to establish credibility before advocating to leave Alonites alone.

Pierre approved of her letter with only a couple of minor suggestions. She sent them to the US's major newspapers: *New York Times, The Washington Post* et al, and even *The Wall Street Journal*. She knew papers wouldn't publish letters already published in other papers but thought if her letter were accepted in one, she'd immediately contact the others.

She also knew her title would be controversial, good for some papers, probably not for the *Wall Street Journal*.

The Fossil Fuel Industry Still Can't Be Trusted

Now that everyone agrees global warming is real and destructive, there is no doubt that in order for Earth to maintain habitable areas, we must continue to make radical changes. Further, we must seek more ways to make currently inhabitable places habitable. While the fossil fuel industry publicly agrees and claims to be working to assist in the solutions, they are continuing dangerous practices and keeping them undercover. On one hand, they support climate mitigation bills while suggesting amendments they can exploit to keep doing business as usual. On the other hand, they donate to members of Congress to defeat these bills. These practices began decades ago, and they haven't stopped. They've merely found ways to greenwash and hide their perfidy.

I am a marine biologist whose PhD thesis is subtitled How the foxes convinced the chickens to help them solve the problem of hens disappearing from the chicken house. Not every company in the industry is culpable, but my thesis names those who are.

Further, the burden of their polluting ways is imposed upon the weakest members of our society. The processing of fossil fuels by oil refineries, the pulverization of coal, and the like have been situated in the poorest neighborhoods. The expected life spans are correspondingly lower. Executives who have children evidently believe their offspring will be able to afford livable places, but this thinking is short-sighted. As the livable places shrink, the quality of life of all of us will be affected. Imagine riots galore. Further, efforts to go 100% EVs require copious amounts of lithium. Once again, some companies mining for lithium are harming the livability of indigenous populations, and this exploitation must stop. There are areas like the unlivable Salton Sea from which it may be more expensive to extract lithium from areas that the Indigenous depend on. Still, it is a price we must pay and pay for ways to minimize the impact on plants and wildlife at established mines or restore them to their original conditions.

To be fair to the fossil fuel industry, they are not alone. Instead they are part of the practice of the wealthy and the opportunists to seek more and more wealth at the expense of ordinary people. Another example is related to small farmers. The wealthy have purchased land and paid people to farm the land. Then they work to pass farm bills pretending they support real farmers, but instead, subsidies are paid to the owners of the land, not those who do the work. Seed companies that once sold seed corn, for example, now do not allow the purchasers to use some of the harvested corn for seeding next year's crop unless they pay again. They can test the DNA of the seeds to prove that the seeds' ancestors came from their product. At one time they designed seeds that produced sterile offspring. Unfortunately, some was sold to farmers in poor countries whose practice was to save seed from year to year. Of course, they were not informed about this aspect of the purchased seeds. In addition, some of the many herbicides and pesticides sold to stop weeds and pests were harmful to humans as well.

I call on the general public and members of Congress to wake up, to step up, and stop these practices. The future of the human population is at stake.

Chapter 56

(Elizabeth & Pierre explore the wonders of NYC)

The letter was sent off, and she and Pierre visited the Metropolitan Museum of Art, Coney Island, the Statue of Liberty, and most of the must-sees in NYC over the next few days. The half-days experiencing what the city had to offer turned out to be a perfect amount to avoid feeling drained at day's end.

Elizabeth spent the rest of her time making appointments with all of the members of Command Center, intending to approach Jackson last. In addition, she made appointments with a member of the House, who has an office in NYC, and a member of the Senate, who is a two-hour bus ride away. Each meeting taught her how to better maintain her cool, make her points quickly, and establish rapport. She understood those she spoke with wanted to appear sympathetic, but that didn't mean they'd support her plea to abandon an attempt to capture an Alonite for study.

She wished she could count on further interactions with Alon as she might garner enough information to satisfy the curiosity of the scientists and the public about another sentient kind of being. She also understood the need to know if there would be any consequences to humans living in submarine cities not far from another civilization.

However, she argued that examining a single Alonite would not answer the question of coexistence.

She and Pierre enjoyed dinner together every night until he was satisfied that he'd seen enough highlights of NYC. They had ventured out to other restaurants but found dining in the hotel like eating *at home* after a long day. They never spoke of the increase in romantic attraction they felt given Rule # 9.

On Pierre's last day in the city, they spent the morning in Central Park, having experienced only a taste of the park at what now seemed months ago. They splurged on a horse and carriage tour and were able to make advance reservations for an early dinner at Tavern on the Green. They dressed for a cloudy day with low wind speed. They saw Wollman Rink, Belvedere Castle, Strawberry Fields, and Bethesda Fountain, along with numerous chirping birds and scampering squirrels. You would not know that Earth was in crisis mode. It was a perfect day for ice skating, and they smiled at couples dancing together along with scared-looking toddlers with legs spread so wide that they looked like Thanksgiving turkey wishbones. Pierre observed that Yellow Submarine was not the only thing named for a Beatles song when they drove through Strawberry Fields. Bethesda Fountain overlooked the southern shore of the Lake in the Park. They learned that the Angel of the Waters statue atop the Fountain is the 1860s masterpiece of lesbian sculptor Emma Stebbins. It was the earliest public artwork by a woman in New York City.

The carriage ride was relaxing physically and fell just short of sensory overload mentally, just the kind of day Elizabeth and Pierre needed. They had cocktails before their dinner at the Tavern. Elizabeth ordered a dish where she had the chef hold the baby octopus but loved the remaining combination of Orzo, Sundried Tomato, Feta Cheese, Kalamata Olives, and Roasted Lemon Vinaigrette.

Pierre had heirloom grain and roasted vegetables. They split an order of Cherry Crème Brule for dessert. Elizabeth noted the dishes in a diary. The food had well lived up to the Tavern's reputation, and Elizabeth said they had earned the right to haughtily say, "I dined at *The Tavern*" like a real New Yorker.

Pierre laughed and took her hand as they strolled to the subway, most content with their fulfilling day. Even the crowded, noisy subway ride wasn't enough to quell their appreciation of a day to remember.

Chapter 57

(The wonderful day is not yet over)

As usual, Pierre walked Elizabeth to her door. She lifted her cheek for his usual peck. Instead, he clasped his hands on both of her cheeks. "Are you someone who believes in meeting the spirit of the law as well as the letter of the law, as I think you Americans say?"

"Why do you ask?" a puzzled Elizabeth asked.

"You know rule #9 of our guidebooks?"

Elizabeth's emotions tumbled in her brain. "Of course. Are you asking what I think you are?"

"I am." He locked eyes with her until she felt she might melt away like the witch in *The Wizard of Oz*. "Well, generally, I do believe in obeying the spirit of the law." She took a deep breath. "But maybe tonight I could make an exception."

"So, I may come in to cap a perfect day?"

Elizabeth unlocked her door and spread the door wide. Pierre took her hand and led her to her bed, putting his right arm around her and sweeping away the bedspread and blanket with his left. He gently lowered her to the bedside, his eyes telling her to make room for him. He gently removed her shoes and his own while not taking his eyes from her face. Once again, he took her cheeks in his hand and began a kiss that both seemed never to end and which Elizabeth never wanted to end. It was like one's first taste of tiramisu, a taste worthy of a lifetime but also caused one to want more. The kisses continued of all kinds: hungry, short, fast as if they were fish, light on the tongue, and stopping occasionally to stare into each other's eyes. On one of the pauses, Pierre began to unbutton her blouse. She looked at him with an obvious question on her face.

"Don't worry. I know I can stop. Wait a moment." He picked up and fiddled with her alarm clock/radio. "I've set it for forty-five minutes. Then I'll leave."

"A lot can happen in forty-five minutes," Elizabeth warned.

"I know, but the letter of the law won't be broken. Now, where was I? Did you know nipples are made for nibbling."

Elizabeth laughed. "Oh, they are, are they?"

"Oh, yes. Did your mother not tell you?"

"MY mother!? Neeever."

"Well, my father told me."

While not taking his eyes off her face, he continued to unbutton, kissing her while slipping his arm under her chest and skillfully undid her bra.

The kisses never stopped as the straps were pulled aside. He began to alternate tickling her two nipple tips with his tongue until they stood at attention, followed by swirling his tongue around the nipples as if he were enjoying an ice cream cone. Elizabeth felt she had skipped death and gone straight to heaven.

He stopped, momentarily, Elizabeth thought to gather himself. She knew she needed to do the same. She feared letters and spirits of laws were doing the Chacha.

He looked at the clock. "I want us to have one more experience."

Elizabeth said nothing. She did not feel in control of herself.

Pierre put his hand between her thighs, touching the front seam of her panties and fondled her there. He seemed to be searching for something or the right position, then gently rocked his hand back and forth.

Elizabeth closed her eyes and was aware of only one part of her body. Suddenly, there was a burst that she couldn't explain despite her wide-open mouth. Finally, she spoke, "Oh my god."

After glancing at the clock, Pierre again took her cheeks in his hands. "See, look, we have five minutes to spare."

Elizabeth laughed. "I'm not sure how much you got out of this, but ..." She couldn't finish her sentence.

"Oh, don't worry about me. I fared just fine. Now, let's see. I understand you are uncertain about the new location for Yellow Submarine."

"Huh?"

"Now, if someone discovers I entered your room, we can say that we talked about Yellow Submarine."

"You've thought of everything."

They rose from the bed. Elizabeth walked him to the door. "What time are you leaving tomorrow morning?"

"Too early to eat breakfast with you. Besides, I have to do laundry."

They had one last kiss and said good night.

Elizabeth had never been so relaxed, but she suddenly laughed about his having to do laundry.

Chapter 58

(Elizabeth's campaign begins)

She spent the next day half-dazed. Fortunately, each interview with a Command Center member went much like the first one, where she was warmly welcomed and praised for discovering another sentient being. "Not only sentient," she would respond, "but highly intelligent. Their species is advanced enough to have invented something akin to our bathyscaphes and has been cursed with corruption in government." That usually brought a laugh followed by a comment about how much information she had gathered, an opening Elizabeth loved. "Yes, and the information has been gathered in the best possible way. I know their power to pass on information telepathically is most interesting, and I understand it would be a major achievement if we could learn to do the same. However, part of me is concerned about the feasibility. We initially thought social media was a huge accomplishment, but look how it came to be used. Sorry, I'm afraid I've digressed from the reason I'm here. I do not believe that capturing Alon or one of his kind will successfully uncover their ability, and further, it's absolutely inhumane. What could be done to attempt to study the inside of an Alonite's brain that isn't invasive, if not risky? Further, it all has to be done in ocean water, so how does that work with air-breathing doctors? Robots? I think

Alon's ability is much like our ability to make intelligible sounds that communicate ideas to one another. That's pretty amazing, but if you took the brain from a deceased person, how would anyone discover how they make these sounds?"

While this analogy seemed to work, the Command Center member would waffle with something like, "I think we will have to wait and see what Director Jackson will come up with."

"Have you heard any of his ideas?"

"Not yet, but there's time."

"Can I count on you to vote no if he doesn't come up with a sound plan to accomplish studying an Alonite that is not invasive?"

"Well, I'm not sure. I'd have to know more."

"And now *you're* being evasive," she would say with a sprightly laugh intended to make her point with the least possible confrontation.

The member would smile, and after Elizabeth requested that they give what she said careful thought, she thanked the member for his or her time.

It was the best she could do. She only hoped her opinions would take root…

Chapter 59

(Meetings with Congress members)

The day after her second-to-last interview, Elizabeth heard from The Wall Street Journal that her letter to the editor had been accepted. Of all newspapers, this was the last paper she expected because of their business orientation. She could only speculate that her letter poked a beehive of opposing responses since all media loves controversy. The arguments caused such a stir that the NYTimes and the Washington Post both picked up on the controversy.

She wondered if it affected the ease with which she could now make appointments with Congress creatures, as she called them. The fact that the term Alonites had replaced 'sea creatures' in the public eye made her feel she had won the lottery. She well understood the power of language. Never mind Shakespeare. Call a rose a gobrog, and it would *not* smell as sweet.

Meetings with a member of Congress typically began with a warm welcome, an offer of a beverage, a thank you for her interest, an understanding that she was in the limelight, and the question of what he or she could do for her. They had been given the purpose of the meeting per the request of the staff. Elizabeth would jovially say, "I have an unusual request, one that does not ask for government money but, in fact, one that is asking you not to approve money to be spent on the Command Center's director's idea to investigate Alonites." From there, the discussion was much like the ones with the Command Center's members, namely that she argued her interactions and future interactions with Alon were the best way of learning about Alonites. She argued that any thought that brain folds could reveal anything about telepathic powers would be like dissecting a finger to see how it could power a touch screen.

She felt better about these interviews than those with Command Center members but still worried whether she was taking on more than she could accomplish.

Chapter 60

(Elizabeth gets a break from an unexpected source)

A producer from the popular Amanda Sanchez's daytime show *What's Up* called Elizabeth after tracking her down through the Command Center. "We'd love to have you on Amanda's show. This submarine thing looks like a popular topic, and from the publicity you're getting, you're just the person for Amanda to interview. One thing, it is critical that you wear the uniform I understand all of the scientists wear."

Elizabeth gulped. "I didn't bring one with me. When that's the only thing I've worn for a year, leaving it behind felt like getting out of jail."

"Surely, the Command Center has plenty for new volunteers, and they must wear out, or people gain or lose weight, so they need replacements."

"I'll check," Elizabeth said but had serious doubts. "When were you thinking of me being on?"

"Let me know as soon as you have your uniform," she said and gave her the number to call.

Elizabeth wasted no time heading directly to Command Center. She explained what she wanted to the main receptionist, who said there were a few on hand, but he'd need Director Jackson's approval to give or loan her one.

He punched a button on his intercom, and Elizabeth heard him say, "Send her in," after hearing the request.

She was grateful she had contained her blunt ways with him so far. Evidently, no member of the Command Center had informed him about her meeting with them as a good sign. Indeed, Director Jackson appeared pleased to see her. "I hear you're becoming quite the rage. That's great for our submarine project. It helps with the bitching about the money being spent as if we are researching how tigers get their stripes instead of the survival of the human species. Raj said you need a uniform for the *What's Up* show. That's a coup for our side."

Elizabeth merely smiled. "Yes, it certainly is."

"What size do you wear?"

"Women's medium. Small will also probably work. I've lost weight working on the submarine."

If Raj can't find one, we'll place a special order and get one to you within two days. He buzzed Raj to get on it.

"Take a seat. I appreciate your dedication and spending so much time on your break on our ultimate cause. The American people are so short-sided sometimes. Help yourself to some coffee from the bureau over there."

"Thanks. I did get to see relatives, although it wasn't under good circumstances. My mother passed away."

"Had she been ill?" Jackson asked in a sympathetic tone.

"She's had severe dementia for four or five years, the kind that causes people to say of her death 'it was a blessing.'"

Jackson nodded in understanding. Elizabeth realized that under different circumstances, she might consider him a nice guy. Certainly, given the big picture of making life on Earth livable for the human population, they agreed. One thing that bothered her, really bothered her, is that it was likely that there would be ways and means for the wealthy to continue to live normal lives, but that life for everyone else would be next to unbearable. That gap has existed for centuries, if not the history of the human race, but it's been widening over time. She also believed that if climate disasters destroyed too many people's lives, it couldn't help but impact the upper class.

Raj tapped on the door and entered at the director's bidding. "I found a woman's small."

"Great," Jackson said, then pointed at a door in a corner behind him. "That's my private closet. You can try it on in there, but come out and let me see how it fits." When she returned, he nodded. "It looks great on you. We're in luck. Get yourself on that show ASAP. The publicity will be great!"

Elizabeth changed and practically danced back to the subway. She should have been tired, but things had gone so well recently that with her tension drained, she felt so relaxed that it felt as if she'd had a good night's sleep.

Chapter 61

(Elizabeth on the What's Up *Show)*

With interviews involving people in power accomplished, Elizabeth could take a break for the two days before the *What's Up* show. Her mom used to watch it, and she tried to remember what it was like. She recalled that it wasn't a program that employed gotchas but was primarily devoid of much content, a fluff show. She might have to work to get some serious points across. She did her thinking in the hotel spa and treated herself to a mani-pedi and a whole-body massage.

When the day came, she arrived at the studio two hours early for makeup and hair. She almost didn't recognize herself, but she didn't think the vast improvement was worth the two hours of work.

Amanda gave her typical dramatic background speech. "Our next guest lives in a yellow submarine. I kid you not. Most of you are too young to remember the Beatles, but they ruled the music world of the 1960s, and those who loved them played their music for decades more. Yellow Submarine was one of their hits. Can you guess where our next guest lives? Yes, on Yellow Submarine, the city in a submarine a mile under the surface of the Indian Ocean. Elizabeth Gayer is one of its volunteers. Someday, many of us may be living in such a submarine." After more yammering, she finally said, "Let's give a big hand to our intrepid volunteer." While the clapping was reasonably loud, television would vigorously enhance it.

Elizabeth fought to keep her lips in a smile rather than a smirk. She sat in a comfy chair opposite Amanda.

"So, first, tell us about your outfit. Is that what you wear every day on Yellow Submarine?"

"Yes, well, not the very same, but an identical coverall. Everyone has only three, so we're washing them frequently."

"How's the food down there? Or rather, how do you get it?"

"We have regular deliveries for some items, but we have a greenhouse that grows fresh vegetables, lettuce, and herbs."

The show continued, interrupted by long commercials with Elizabeth responding to questions: yes, we have a dentist, doctor, laundromat, movies, library, etc., etc. She had to force a comment to an unasked question in describing the walkway where one could observe fish swimming alongside.

Amanda liked that. "Still," she said, "how do you spend your time? There doesn't seem like much to do."

Finally, an opportunity to describe explorations in bathyscaphes.

Amanda's eyes widened. "You should have told us that sooner. That sounds amazing. We're almost out of time. So, you're like steering your own little submarine?"

"Yessss."

"That sounds really exciting, but I want to get to one final question. How can our public get coveralls like yours?"

"I have no clue. Command Center contracts them from some company."

"Get back to me on that, and come on folks, another big hand for Elizabeth Gayer, who lives on a submarine."

Elizabeth walked out feeling like her eyeballs could leap from their sockets and roll across the room. Jackson would be pleased with the publicity, but she hadn't been aggressive enough to squeeze in the issue of Alon.

Elizabeth had to take several different subways to return. The station had picked her up in a limousine, but now they were done with her. She couldn't decide if the publicity was good for her cause or had just now labeled her as a lightweight.

Chapter 62

(Director Jackson summons Elizabeth)

Director Jackson left a message at the hotel for Elizabeth to call him. Elizabeth feared the worst. Part of her campaign depended on Don Jackson not being aware of the strength of her opposition for as long as possible. To her relief, It turned out to be a request for her to return to Yellow Submarine in its new location and conduct an exploration. "It's a new territory, and I thought you might like to be the first to explore the area. Besides, you have established some rapport with this sea creature you call Alon. Maybe you'll be lucky enough to find out more about him. There's a delivery scheduled for tomorrow and another in two days because we have to replace some equipment. You could come back in two days. Please don't feel pressured to lose two days of your break. Command Center would appreciate it, but it's certainly not necessary."

His graciousness amused Elizabeth. No wonder he was elected director. "I'd be happy to check out the new territory. I was looking forward to seeing some new surroundings and possibly new sea life. As for Alon, I'm glad you understand that direct communication with Alon is the best way to learn about his kind."

Silence. Then Jackson said. "Be at the dock by 7:30 tomorrow morning."

She refused to be dismissed and said, "On another note, I assume you watched What's Up? I think Command Center could make money selling our uniforms to the public. Amanda Sanchez has quite a following."

"Thanks for reminding me; I was thinking the same thing. We could even have some made for children."

"Oh, my sister called me and wants one for her teenage daughter. Any chance of getting a woman's small soon?"

"Yes, I've requested a shipment of fifty more uniforms to see how it goes. I'll reserve a small for your niece. In fact, give Raj her name and address, and we'll send it from here. Consider it payment for your overtime."

Damn, Jackson was sounding too pleasant and agreeable. It would be impossible to discredit his dreadful position on Alon as his typical off-the-wall behavior. She reminded herself of the adage of honey being sweeter than vinegar, and decided to attempt to convince rather than condemn.

Still, Elizabeth left Command Center with a positive feeling. She'd accomplished what she wanted before going full-on public about her opposition to capturing and studying Alon's species. She'd completed all of the appointments with all of the influencers. She had a wonderful time in New York City with Pierre, and she looked forward to a jaunt with Tyrone to investigate new territory.

Chapter 63

(Elizabeth returns to Yellow Submarine)

As the only passenger, Elizabeth enjoyed chatting with the pilot and one engineer on the trip back to Yellow Submarine. She attempted to sound out their views on capturing one of Alon's species for study without revealing how strongly opposed she was. They hadn't thought about it but considered it a no-brainer that she was right once she fully described the situation.

They moved on to how lucky they were to participate in this experimental venture. The engineer said, "Things seem to be going swimmingly." Elizabeth smiled, thinking he'd get along beautifully with Hans. The major unknown was whether people would experience mental issues after years spent in a finite space. When they embarked, Yusuf was waiting for the delivery as Sanjay usually did. The shock on his face surprised Elizabeth. "You didn't know I was coming?"

The pilot of the delivery submersible began to laugh. "Everything was last minute, and we thought it would be fun to keep it quiet."

"I knew you were dedicated, but I did not think you would foolishly shorten your break. It really is not a good idea."

"Only for two days. You know that Command Center wants to upgrade some of the equipment. They'll load the old and return with the new in two days. Director Jackson seemed to think I should be with Tyrone for the first exploration at our new site."

"I see, but he never mentioned it to me. As team leader, I should have been informed." His deep frown revealed his annoyance. "It is too late to explore today. You will have to wait for tomorrow. I will inform the team over the intercom and ask Alessandra to prepare a special meal."

"Oh, she's taken the place of Hans?" Elizabeth said, but it wasn't a real question.

"And she has been adequate," Yusuf said. "I've assigned Heather as her sous- chef."

"It seems as if I've been gone longer than I have been. My mother died, and I've spent time with family."

"You have my condolences."

"Thanks." She didn't feel like discussing it with him and changed the subject. "I've been working hard on a campaign … uh, to promote our good work here." She had begun to say something she'd regret but caught herself.

"Excellent! What have you done specifically?"

"I was on a television show that discussed submarine life."

"Ah, most exceptional. You can say more at a team meeting tonight. I must go." He nodded and left, to Elizabeth's relief. She headed for her room to change and check out the nursery.

She met Sidney in the nursery, who looked surprised to see her. "You aren't due back," he said.

"Director Jackson wanted me to explore the new territory first. Then I'll finish my leave." She looked around. "All the plants look bigger and healthy. They must like you. "

Sidney smiled. "I don't know about that, but I am pleased with how things are growing. How's your break going?"

"It's a long story. Yusuf is calling for a team meeting tonight. He'll ask me to give a report, so besides my mother's death, I'll save the rest."

"Oh, I'm so sorry your mother died." He hugged Elizabeth.

"Thanks. It was expected." Now, she dreaded all the sympathy that would be shed on her, almost wishing she'd not mentioned it. "How have things been going down here with Yusuf in charge?" She walked slowly through the plant aisles.

"Oh, I suspect you can imagine. He's his usual hard-to-take self, but he is tending to everything required." He nodded and raised his shoulders, indicating it was the best one could hope for.

"You must be due for a break soon." She popped an irresistible strawberry into her mouth.

"Just a few more weeks. I'm looking forward to bringing back some edible succulents for the nursery."

Elizabeth's jaw dropped. "I didn't know such things existed."

"Yes, some cacti, dragon fruit, and aloe, to name a few. They're quite tasty, and, like all succulents need little water. It's something the entire world should be looking at. The other thing that would be great food-wise is snakes. Yellow Submarine should carve out a room for them."

Elizabeth couldn't believe it. "Really? Have you talked to Yusuf about this?"

"No, I thought I'd wait for Sanjay and Pierre to return." Sidney shrugged.

"Why wait? I'd support you. We're meeting tonight." She gestured with her hands.

"I'll think about it." He wagged his head from side to side.

Elizabeth left to see if she could find Tyrone. Sanjay used to post a schedule, so she headed to the situation room. It turned out he had already altered tomorrow's schedule to include her. She had time to get in a walk on the top floor.

She didn't see a new species but enjoyed seeing more diversity of kinds of anglerfish. Orange ones surprised her, and one with a headlight at the end of its esca — the little pole that lures smaller fish. Elizabeth liked to call them handles. Despite a PhD in Marine Biology, she had never heard of a type with headlights. She did remember thinking it odd when she learned in graduate school that they were carnivorous, and even cannibals.

Walking and watching, Elizabeth felt she was home. Score one point for submarine livability. Of course, she had been spending her time in a hotel, so perhaps the comparison didn't accurately reflect how she'd feel if she had been in a place of her own.

It felt even more like home when she ate dinner with her colleagues. Not surprisingly, they said that once Yellow Submarine anchored, it was as if they'd never moved. Everyone expressed their condolences to Elizabeth. She would quickly change the subject and advise people to spend some of their break time in NYC and describe her favorite sites.

Despite enjoying her comeback immensely, Elizabeth's energy began to fade, and she begged off for a good night's sleep. It had been a long day. However, it didn't turn out peaceful. She had a nightmare where Jackson was sawing open Alon's head.

Chapter 64

(Elizabeth and Tyrone's encounter)

Bombarded by her teammates at dinner, Elizabeth had to postpone talking to Tyrone and realized she missed him. The next morning, before climbing on board the caterpillar, she felt as excited as a young child looking forward to seeing Mommy come home from the hospital with a new baby sister.. For the first hour, besides anglerfish, she saw several fish she'd only seen in textbooks. The batfish looked like a skinned baby pig that had been flattened. Many others appeared to have a translucent quality, such as a snailfish. One exception was the pitch-black pelican with a beaked mouth and tapering tail. Tyrone seemed impressed she remembered the names.

When they hadn't seen any fish for some time, new or familiar, Elizabeth trumped up a question for Tyrone. "When you were a kid, where did you do your homework?"

"At the kitchen table. We were a kitchen table family. Dad paid the bills there. Mom wrote letters there. Often, all three of us at the same time. Our place was too small to do anything else."

"I can picture you there," Elizabeth said, smiling. "Who called the shots in your family, Mom or Dad?"

Tyrone had to think. "I'd say mostly Mom, but it depended on the issue. Since Dad paid the bills, he'd give orders about leaving lights on, turning up the furnace too high, whether we could afford this or that. Mom made the social calls: which church services we attended, who we visited, and what potluck events we attended. If anyone was sick or in the hospital, we needed to visit with an appropriate gift. Usually, homemade cookies."

Elizabeth found Tyrone's details charming. His family sounded typical and healthy, so whatever experience suppressed Tyrone's comfort level around Whites had to happen in young adulthood. "What things did they argue about?" she asked.

"They didn't really argue that much, but mostly about me, what I could do or wear. Then they could get into some doozies." Tyrone's tone of voice suggested amusement, not pain. "Mom was cautious, and Dad would argue that I couldn't grow up afraid of the world. When he lectured me about my behavior, he'd look me in the eye and ask, 'We straight now?'" Mom took for granted I'd behave to her liking, but if I didn't, she'd guilt trip me. She'd say, 'Look what you did to disgrace your poor mama, y'hear?'"

"Does that mean your dad didn't warn you about not looking like you were going to steal a candy bar like your mom did?" Elizabeth asked.

"He thought my mom overreacted about that, but Dad told me that it's not enough to do the right thing. It was important for people to *know* that you did what was right. That's advice for everyone. He said that the chances of being abused by the police would get worse as I got older and advised me always to defend myself but stay calm. He said, 'If need be, tell them your mom has been in the office of the President of the University of Chicago and that they are on a first-name basis.' Mom laughed at this advice.

Mom couldn't be any more cautious. She always said, 'If in doubt, don't do it or don't say it.' Dad didn't like the idea of me kowtowing to anyone, but I should be respectful and non-confrontational but innocently ask a Whitey to give reasons for being stopped or questioned. In general, if I were going somewhere, they'd both tell me to behave. On the other hand, if I were mistreated, it would be Mom who would get on her broom and fly to the clerk or teacher who done me wrong and let them have 'what for.'"

Elizabeth's lips curved up. "I think I'd love both of …" Suddenly she yelled, "Stop, I see Alon. Oh my gawd, he found our new location!"

Tyrone slowed and waited for Alon to come alongside. Elizabeth couldn't believe it.

Soon, the bathyscaphes were side by side. Alon wore a big grin at Elizabeth's surprised face. "I see you, no expect to see me again," he telemessaged.

Elizabeth nodded with an equally sized smile. She noticed he had a new pilot.

"I follow your submarine move," he conveyed as if he had read her mind.

"How you like I teach you to telemessage?"

Elizabeth's excitement answered the question.

"No, get excited. Not know possible. Only few my kind can send messages. I learn, not easy. It take time and practice. Then you different species, but we so much alike."

Elizabeth found this information amazing. The ability to telemessage was what the director wanted most of all. This new information felt like the Christmas she had begged for a bicycle to discover what Alon had in mind.

Alon telemessaged, "Ready?" When she nodded, he messaged, "Face me, close eyes, and clear mind. Slowly imagine some words leaving your brain over and over. I wait."

Elizabeth calmed herself, turned toward Alon, closed her eyes, and once feeling peaceful, tried to envision 'hello' leaving her brain over and over. She tried not to expect Alon to message back but couldn't help it.

After ten minutes, Alon messaged. "You can stop now. Hard to do on first effort, we try another time. I think I tell you as much as you might want know about my world, but I like know more about yours. I learn from ask yes - no, but I think you like do better than that."

Elizabeth nodded vigorously and rued the fact that she couldn't come back the next day as she watched him leave. While dubious, the idea that she might be able to telemessage turned her brain abuzz. Then she received a last telemessage. "Not continue at this altitude in direction you headed. Rogue currents possible."

She had ignored Tyrone in order to concentrate. Now, he returned to her. "What just happened?" After Elizabeth explained, his eyes opened wide. "That would be a blessing from God, as my mom would say."

"So that's what your mom said a lot?"

"Yup, maam," he said in his child character.

Elizabeth got it. "Well, I seriously doubt I can telemessage, but I'll give it my best effort. Where were we before Alon surprised us?"

"I think you were about to say you'd love both of my parents."

"You're right. I'm sure I would. Oh, Alon said we should turn around; the way we are going is subject to rogue currents."

Chapter 65

(The caterpillar is in trouble)

After Elizabeth passed on Alon's warning. Tyrone barked. "Now you tell me!" As if she had called trouble's name, the caterpillar lurched to the left while Tyrone held tight to the controls. Straining his muscles to their limit, he could not hold the caterpillar to its charted course.

"What's going on? Can I do something?"

"We hit a current from hell. Sit tight."

Elizabeth did just that. She knew better than to interfere. She couldn't help but wonder if the new site was ill-chosen or whether this was a random current. It pained her to watch Tyrone struggle with the controls. All kinds of fish were bashed against the scaphe. Seconds seemed like minutes, and minutes seemed like hours. The scaphe was being propelled at an unbelievable speed, thrusting them up, crashing them down, and shifting right and left in random order. At one point, they flipped over completely. Tyrone was strapped, but Elizabeth tumbled slowly onto the right side window, hitting her head and then back. She struggled to strap up while holding one hand over her head for protection. Tyrone grabbed her at one point so she could use both hands. Suddenly, they stopped being tossed and coasted in an unknown direction. Elizabeth breathed a deep sigh of relief. "Oh, my god!"

"Don't relax yet," Tyrone said as he struggled with the steering mechanism and tried to reset the controls. "I'm afraid we've been driven out of range of Yellow Submarine, and I don't know how badly our controls have been damaged."

What does that mean if they are? Please don't say we're dead in the water."

"All right, I won't say it, but I don't know how we'll get back if the controls aren't working."

Elizabeth knew they'd initially lurched to the left, but it was impossible to remember their tumbleweed path after that. She mentally kicked herself for not telling Tyrone Alon's message immediately after their encounter. They might have been able to change course in time to miss the damned current.

She opened her mouth to ask Tyrone how it was looking but stopped. He needed to work without her interruption. She silently repeated "please, please, please" in her brain. After the longest ten minutes of her life, she heard Tyrone sigh. His lips tightened as he turned his face toward Elizabeth and shook his head. "The worst thing is that the coordinate gauge is not displaying. We're at the mercy of the currents. We have to wait for help and hope we're not out of range of the submarine's ability to track us." He paused. "And hope our oxygen lasts long enough for them to get to us."

Elizabeth knew better than to panic and gobble more air. Haru and Yusuf could pilot a rescue scaphe, but she wondered if anyone knew they were lost. Anyone in the situation room would soon realize that they were off course. 'Sit tight' was all they could do.

"Are you OK?" Tyrone asked.

"I will be, when we're back on Yellow Submarine." She knew that keeping calm could improve their chance of survival, that if anyone could salvage the situation, it was Tyrone, and most of all, he didn't need her fear to deal with. "Don't worry about me. Just see if you can get the controls back in working order." She looked at her watch. "We should have forty-five minutes, right?"

"Sounds about right, depending on how long it will take us to return."

They were silent, conserving air for the next half hour, while Tyrone worked frantically but got nowhere. "It doesn't look good," he said, "I've released pellets so we'll drift higher, which we would have to do at some point to get back. That's assuming my sense that the current hasn't lifted us so high we're above the submarine. The steering propellers are working fine. If only we knew where we were."

"Would it make sense just to go up?" Elizabeth asked.

"Not really; we don't have enough air to get to the surface, and chances of getting near the submarine are next to nothing."

They were silent for a few minutes, which seemed like half a lifetime. Elizabeth could not help but think, "We .. are .. going .. to.. die." "You know, if we are about to spend the last fifteen minutes of our lives, I don't want to sit here like a frog on a lily pad," Elizabeth said. "Let's make the most of it. I'll share something with you I never would otherwise, and if you're willing, you can tell me something you've never shared." She told him that she and Pierre were more than platonic and had come close to breaking the rule about sexual relationships between team members. "We really care about each other, and it's going to make it hard to work together on Yellow Submarine if by some miracle we make it back."

"I was always suspicious of you two. Everybody is. It's a dumb rule. I know you would never let your feelings interfere with what's best for our submarine home." He paused. "You want to know the real reason I don't like telling you about my life or, rather, that you had to drag what you did out of me?"

"You know I do because I care about you."

"I don't trust Whites. I've trusted Whites in the past, but it didn't turn out well. I trusted a white husband and wife, and I got burned. Remember the little girl I told you about when you asked about my school friends?" He continued when Elizabeth nodded. "Astrid was in my class or some of my classes all the way through high school. She was white, but we started dating. I met her parents, and they were great, or so I thought. After graduation, I stayed in Chicago, going to the University of Chicago, and she went to the University of Illinois. We stayed connected, talking twice a week. Urbana was about three hours from Chicago, so she not only came home during breaks, but at least once a month. Of course, when she was home, we continued our relationship. We were madly in love, and we became engaged. I didn't do the *ask-the-dad-for-her-hand thing*. We announced it together." He paused, choking up, and Elizabeth wondered if he could continue. She suspected she knew what he'd share next. "Her parents said nothing, which should have been our first clue. Only when we set the date shortly after we were to graduate from university, did they tell Astrid that she couldn't marry me. When she asked why, they said that she should know. She got them to admit finally that it was because I was black. They hadn't said boo before because they assumed she'd meet someone else since she did date a few other guys, and they wanted to seem non-racial in every way. In fact, they claimed they were non-racial in every other way, but didn't want their grandchildren to suffer from our racist society. Astrid was shocked and insisted we'd marry with or without their blessing. The bottom line is that they eventually wore her down. I vowed never to trust a White again. That's why I don't like talking about my background. I did trust you, but I also trusted Astrid's parents. What a mistake that was. I so often think about how Astrid's parents misled me."

Elizabeth said nothing, trying to absorb Tyrone's pain while hugging him. "Well, you know I don't have a daughter you might fall in love with." She was surprised she could be so flip when they had little time to live. This wasn't a movie or a book where characters were put into perilous situations with little time until disaster would strike, but at the very last minute, rescuers like the cavalry would come rushing in. She didn't say she knew they were going to die. They both understood it. Their oxygen had to be dangerously low. Time had become most strange in her perception. It had both stood still as the harrowing thoughts crept through her consciousness, but it had sped by regarding the amount of time remaining. She started to say something when Tyrone put his finger to his lips, warning her to be quiet.

She whispered, "What?"

I thought I heard a squawking sound. He headed for the microphone. "It's just static, but …" He stopped and appeared to be trying to make sense of what he was hearing. "I think MAV — Mechanically Augmented Voice, a revolutionary transmission of code from a combination of radio and acoustic signals to human language — may be trying to get through to us." Elizabeth joined him in trying to fathom words out of the squawks. Please, please, please echoed through her brain. Suddenly, they felt a jolt stronger than the current they'd thought they'd passed through. Could a whale have hit them? They began scanning their surroundings. No whale in sight. "I think we're being dragged," Tyrone said with wide eyes. "Look for a chain. We may have been found!"

Elizabeth stood close to the thick window in the direction they were traveling. She looked up. "We are chained.!" Tears began to stream down her cheeks. They've got us!"

A long minute later, they could receive communication again, and their controls began mysteriously working. They heard MAV. "We have you and are hauling you back. You should regain the ability to communicate soon. Let us know ASAP whether your controls are working. You'll get back faster if you are self-propelling, and heaven and hell know you can't have much oxygen left." Tyrone flew into action, running tests on their controls and notifying whoever was in control of communications of their status. The most important controls were the ability to dock properly with Yellow Submarine and the ability to steer. However, they could not tell if the docking mechanism would work, and the steering response time was slowed. Time was of the essence. They could take the shortest route to Yellow Submarine but still not survive.

Suddenly, Tyrone slapped his head. "Oh my god, I'm an idiot! I forgot about our diving suits and their oxygen tanks. We're going to need them."

Elizabeth grabbed two tanks from the locker behind her. There were four in all, as four was the maximum capacity of the caterpillar. Even though they were smaller than the standard size for divers, there might be just enough oxygen to get back. They had drilled donning the suits as quickly as possible, but Elizabeth hoped they would not need the suits since the docking mechanism display indicated it was functioning normally. However, Tyrone insisted on preparing for the worst. "We don't want to die because of the time it takes to get these rubber pajamas on." Despite establishing the path to survival, Elizabeth and Tyrone had no clue whether they would make it in time. Neither of them remembered exactly how much oxygen the tanks contained, but they knew enough to know the time to get back was dangerously close to the limit.

They learned Haru was piloting the rescue vehicle with Haihong. When Haru told them they were not far from Yellow Submarine's dock, they had depleted two of the four tanks and were on the last two. They indicated the obvious that Tyrone and Elizabeth should dock first. They could abandon their caterpillar and swim in if their docking mechanism failed at the last minute. Elizabeth's nerves were nearly fried, but she dwelt as much as possible on the fact that the caterpillar appeared in shape for docking. When they bounced into the dock, Elizabeth exhaled a deep breath she didn't know she was holding. The doors to the exterior closed, and the sea water was sucked out, a routine procedure Elizabeth and Tyrone had experienced multiple times, but when the last drop of water left the area, she began to cry. When the doors of the entrance to Home Sweet Home of Yellow Submarine opened, their teammates stood ready to greet them with cheers. They closed the docking doors, dropped their oxygen tanks, and peeled off their suits to the cheers of their comrades. Elizabeth's tears turned into a steady stream. Tyrone also had tears in his eyes.

Heather hugged both of them. "Whatever you two want, a massage, a good meal, sleep, you've got it! Later, you can tell us what happened. You're safe, and that's all that matters." Moments later, Haru and Haihong exited their caterpillar. The entire rest of the team waited. "You must have kept calm and carried on," Heather said, hugging Elizabeth and Tyrone.

"No, we did not keep calm, but we carried on," Elizabeth said. "Tyrone was great."

"She's wrong; she kept me calm," Tyrone said.

Chapter 66

(Post-disaster meeting)

Elizabeth opted for a soaking bath followed by a long nap, while Tyrone needed to stretch his legs. They agreed to tell their story at the evening meal. Out of habit, Tyrone wanted Elizabeth to tell the story. Elizabeth conceded to Tyrone's discomfort speaking to groups. He was such a jewel, and she so wanted him to enjoy the company of good friends. Now she knew the source of his reticence. She couldn't imagine losing a lifetime with someone one loved because of racism. Indeed, how do you trust seemingly good people again?

The team had quickly thrown together a simple chicken and rice dish. Afterward at the informal meeting, Yusuf asked Elizabeth and Tyrone to tell what happened.

"Elizabeth's got this." Tyrone didn't know that she, in fact, did.

"There's not much to tell, really," Elizabeth began. "We hit a current beyond Caterpillar's ability to control, and it lasted so long and tossed us about so violently that we ended up out of range of our *home*," she said with a sad emphasis on home. Our communication wasn't working, our controls weren't working, and we didn't know our coordinates. We were able to steer but it was pointless to wander from our position. Tyrone, tell everyone what it felt like, and how frightened you were. I know I was."

Tyrone blinked and gave Elizabeth a quick look of displeasure. "I was thinking that we wouldn't be in such danger if we hadn't changed locations. I suppose that a random current can always happen, but we had none in our old location. Elizabeth was awesome. She didn't seem scared. That helped me a lot. To save oxygen, we didn't talk for a long time. We had given up when we decided to share something we had never told anyone instead of just waiting to die. But don't ask us what we shared because neither of us will tell, right Elizabeth?" He arched his eyebrows and tipped his head toward her.

"Right," she said, smiling, her heart warming at Tyrone's commentary. He always did just fine speaking to the group so despite his good reason to not trust Whites, any animosity was never evident. "Tell them what we both forgot, but you remembered at the last minute."

"The oxygen canisters that came with diving suits. I can't believe they weren't the first thing I thought of as backup. I must have been more stressed than I thought I was. They contained just enough oxygen for Haru and Haihong to get us home. So, we're still alive." He paused and his face turned impish. "Elizabeth, how did you stay so calm?"

"I faked being calm. I'm still exhausted from the stress, but Tyrone and I are so grateful to everyone who helped get us back. Right, Tyrone?" she said bouncing the ball back at him as if they were in a hot potato contest.

"As my momma would say, 'you betcha,'" Tyrone said, making Elizabeth grin. She hoped upon hope that his sharing their near-death experience would be a turning point, and Tyrone could begin to put his hurtful loss behind him.

Yusuf ended the meeting by announcing, "I am directing a separate meeting regarding our new location. I favor returning to our old one. I know the current Tyrone and Elizabeth encountered is unknown as a regular current, but if we are in an area subject to rogue currents, we need to reconsider this site. I have always said this move was ill-considered, but no one listened to me, of course."

Chapter 67

(Back to NYC to launch her campaign)

On the return trip to NYC, Elizabeth pondered over the plans she had to make. One goal may not need her to do anything. Yellow Submarine had to move back to its original position if scientists were to continue exploration of the deep ocean. Sanjay was recuperating in the hotel that housed team members being trained or on break near Command Center headquarters, and no doubt he will have advocated for the move back. The critical decision she needed to make was whether she should tell Jackson about meeting Alon, and if Alon might, just might be able to teach her how to telemessage. That's what Jackson really wanted for the human brain, but what are the chances she could accomplish such an incredible feat? Without Alon to guide her there'd be no chance, and she couldn't even be certain she'd see him again. She almost wished she hadn't boarded the bathyscaphe back to New York. She was tired and postponed decision-making until she'd experienced a good night's sleep.

Haihong was on the return scaphe, dozing across from Elizabeth. When she snuffled, Elizabeth smiled. She wasn't the only one in need of good sleep. However, her busy brain made it impossible. She hoped that the two could hang out for a bit in New York. Haihong would be staying overnight in their commissioned hotel, and they could have dinner together.

The announcement that they were about to dock woke Haihong. "Get a good nap?" Elizabeth asked.

"Yes, thanks. I wasn't much company."

"Don't worry about it. Want to have dinner together in the hotel tonight?"

"Yes, that would be nice." Haihong looked genuinely pleased.

"I don't know how much time you've scheduled in NYC, but I recommend that you see as much of the city as you can. Even if you've spent time here during training, there are more interesting places to visit than you can see in an entire year. I'm sorry I'll be too busy to show you around, but I can give you ideas. If you're by yourself, it's probably best or at least safest to take tours. Are you going to go home?"

"No, but I have relatives in the Bay area in California that I'm going to spend time with. I knew them when I was little, before they immigrated. They'll suffocate me with food and all kinds of treats and events." Haihong grinned at the thought. "I know they'll consider it dangerous for me to do anything on my own. I'm not sure how I feel about that, but …" She shrugged.

"I guess parents and relatives are the same the world over," Elizabeth said. In their defense, during COVID, Chinese in San Francisco were often attacked physically and verbally in public. These racist crimes made national news. That's changed, but I'm sure your relatives are still taking precautions.

The pilot announced six minutes until arrival and cautioned them not to try to step onto the wharf on their own but to wait for someone to give them a hand. Elizabeth smirked. They'd have to be slithering snakes to avoid the helpers.

They were soon seated next to each other on the bus for the short ride to the hotel. After checking in and setting a time to meet, Elizabeth took a nap in the fluffy-looking bed that welcomed her. She had a wild dream where she and Alon were touring New York City, and she could telemessage. They talked about how alike their societies were, but Alon's sounded worse. She had never seen Alon's lower body, so in her dream, Alon took baby steps on his split fins. The way he described males and females she concluded Alonites weren't as sexist as human males. As dreams often do, this one morphed into a time when she was ten. They had moved into the new house that her dad had practically built himself. They lived in the basement while the upper two floors were being finished. She and Alon were hammering shingles on the roof. She took a big swing at one, lost her balance, and tumbled off the roof. She woke up on the floor of her hotel room. She looked around. It took at least thirty seconds to remember where she was. She looked at the clock. She had five minutes to get ready to meet Haihong. She pulled on some wrinkled clothes and hurried down to the dining room.

Haihong was waiting for her at the entry, and they were soon escorted to Elizabeth's favorite corner table. The server, Marcello, remembered her and commented on her dining with a new friend. "Yes, another friend from Yellow Submarine."

"We appreciate business from Yellow Submarine," Marcello said. The chef made a bigger menu and hired sous-chefs from all over the world. Easy because New Yorkers are from so many countries." As if he suddenly realized he wasn't to engage with customers, he said formally. "Let me tell you tonight's specials. We have an Italian pasta of creamy rollups with marinara sauce, which I can highly recommend. The pasta will melt in your mouth. We have tea-smoked duck with plum sauce and pancakes, a tasty bouillabaisse, and beef bourguignon over your choice of rice or orzo. I'll give you some time." He practically bowed and left. "The food is exceptionally good here," Elizabeth said. "I don't think you can go wrong with anything, but those specials sound especially good. I'm going for the tea-smoked duck."

"I think I don't need to look at the menu either. I'm up for pasta. Everyone thinks I'll go for Chinese food in restaurants, but I once ate it every day, so … American Chinese food has more meat in their dishes than in China."

They placed their order with Marcello's recommended choice of wine, a vodka gimlet for Elizabeth, and a Tsingtao beer for Haihong. "So, you do stick to some of your country's favorites," Elizabeth said with a smile.

Haihong chuckled. "Yes, there's not much choice on the submarine, although Hans does an excellent job. I know how much he enjoys working in the kitchen, so I'm happy Sanjay lets him cook most of the time. I do not like it when Yusuf cooks. He tries to make Turkish food, then says people do not appreciate fine food if they do not praise what he makes." She made air quotes with the word 'fine.'

"Yes, it's pretty clear he's never done any household chores growing up, so thinking he could cook is like thinking he is a genie." Elizabeth snorted, remembering his comment about his colleagues' lack of sophistication.

"He's never wrong but often underappreciated." Even as she enjoyed Haihong's company, she struggled to keep her mind from wandering to the decisions she had to make. Life on the submarine allowed few opportunities for one-on-one serious talks. The tables for four seemed to invite themselves to be filled. She postponed getting Haihong's advice on ideas on how to stop Jackson until dessert or after-dinner drinks. Haihong had common sense.

After the dinner plates were cleared, Elizabeth took a deep breath. "You know how much I oppose capturing Alon or one of his kind for study." Haihong nodded as Elizabeth continued. "I want to launch a campaign against the idea, but you also know how determined Director Jackson is. Rumor has it his family is super wealthy and has donated to, I don't know how many well- funded charitable organizations. In our country, money means power in many ways, especially in influencing government. Sometimes, I think we live in a plutocracy." She shook her head in disbelieving wonderment.

"Is a plutocracy government by the rich?" Haihong asked.

"Yes."

"It seems the rich have too much to say in all governments."

Elizabeth nodded. "You are so right. That's why we need to fight. I want to launch a campaign against interfering with Alonites. A prominent newspaper accepted an opinion piece I submitted against corporations sabotaging efforts to contain climate change." She made a curious face and added, "I've appeared on a fluff TV show where I talked about life on Yellow Submarine."

Haihong frowned. "Fluff?"

"Oh, sorry. We call feathers fluffy because they are soft and airy, so TV shows are called fluff if they are more about entertainment than serious topics. Often, the hosts talk as if they are providing valuable information. Sorry about using one of those words that some famous person uses, and then it becomes part of our vocabulary."

"No, don't be sorry. I want to learn these words."

"At any rate," Elizabeth continued, "I hope I have achieved a bit of credibility regarding the role of Yellow Submarine, enough that my advocating against capturing Alon might be taken seriously." She sighed. "But I need to be careful about opposing Jackson. I plan to make every effort to change his mind before attacking the idea in full force, even though I suspect it's impossible to dissuade him. What I'm wondering is whether I should tell him that Alon is trying to teach me how to telemessage, when I can't imagine I can succeed." As she said this aloud, it occurred to her it could well mean Jackson would delay his search, which was something. In fact, that's exactly what Haihong pointed out in response. Then she remembered. "I just wonder if it would be safer if Jackson didn't know Alon and I are communicating again. I normally make decisions based on logic, but for some odd reason, I want my gut to feel good about what I decide."

Haihong put her hand on Elizabeth's. "I understand your problem. Let me think about it and let's talk tomorrow at breakfast."

"Yes, let's," Elizabeth said, which fit her plan of sleeping on the decision.

Chapter 68

(Elizabeth schemes with Haihong)

Elizabeth woke the next morning to her phone signaling a message at the front desk. Not surprised, she learned Director Jackson wanted to see her — today! She called his office to set up a late afternoon appointment and headed for breakfast. Haihong already seated, waved at her.

"So, now what are your thoughts?" Elizabeth asked.

"I think you should do what you can to make Director Jackson change his mind first. You tell him about how you might learn telemessaging and the value of that for learning their culture. You can tell him that they are like us, and we don't want to be captured. Then, if he disagrees, try arguing telemessaging must be learned and is not a special part of the Alonite brain, and surgery will likely not tell him what he wants. Also, talk about the expense."

Elizabeth nodded. "Yes, I should emphasize that Alon's ability was learned. I definitely do not want to tell him that Alon can telemessage from a distance, or Jackson will think that it's even more important to discover how he does it."

"Sounds like every idea you have has good and bad aspects. " I don't think you should make him think you *can* telemessage. Just say that it is worth a try to find other ways to communicate. If he gives you a chance to try, that would be good."

"Yes, delay would be useful. I can also stall by claiming I succeeded in teaching him a few basic words," Elizabeth mused. " I just hope Alon won't give up on contacting us when the submarine is moved back." She paused for a few seconds. "Alon knows the location of our first site and shouldn't be surprised about a move back since he warned me about the rogue currents."

She looked at the ceiling pondering. "I've visited members of Congress and members of Command Center about my concern. I don't need to be here to work on a final attempt at ads or some other public communication. So, all that's left is to approach Jackson one more time before I go all out. You've helped me sort out a plan."

"I think you would have come up with the same plan without me." Haihong pleasantly smiled and gently nodded her head.

Elizabeth took a deep breath. "Thanks for your confidence in me, but you don't know the relief you've provided. I won't take any more of your time." The entire dining room was full, and Elizabeth noticed a line for breakfast. She marked the subway maps she had picked up from the front desk, underlined the route to Central Park, and circled the numbers of the subways she'd need to take. "Have a great time!"

Elizabeth gathered her thoughts on the way to her room. A strategy was one thing, execution another.

Chapter 69

(Elizabeth meets with Jackson)

After Elizabeth described her close escape from death, Director Jackson said nothing at first. He splayed his fingers together in a steeple with elbows on his mahogany desk and appeared to be taking time to put his head around what happened. "So, do you have any reason to think that this wasn't a rare fluke? Moving back so soon seems burdensome."

Elizabeth was relieved to have another topic to discuss before bringing up Alon. "Well, this rogue current happened on our first trip out, so that seems ominous to me." As a scientist, she assumed Jackson would want caterpillar explorations to continue, particularly because of the encounter with Alon. "Also, I think everyone is spooked, and if people didn't want to explore cramped inside the caterpillar before, they're going to be even more reluctant now."

Saying nothing, Jackson's face revealed that he considered her argument reasonable. Elizabeth gave him time to ponder. Finally, he said, "I see. Then I agree. We move back. We can also suspend explorations for now and assign people to observe from the walking deck."

His use of the pronoun *we* amused her. "That sounds like a great idea to me. To be honest, I am reluctant to venture out from that location unless we stay super close to Yellow

Submarine, and then we may as well just observe from the deck. Yellow Submarine is not going away any time soon, so taking a break in exploring isn't going to matter much."

Jackson nodded. "OK, I'll let The ROE team know." He began to rearrange papers, which Elizabeth took to mean he felt their discussion was over.

She took a deep breath. "Before I go, may I ask if you still want to capture one of Alon's species?"

Jackson stiffened. "I think that for the sake of progress in science, it's almost a given that we must do all we can to discover how they telemessage. You say they are so much like us."

"Remember the fable of the goose who laid golden eggs?" Elizabeth asked unplanned and coming out of nowhere. "The farmer became impatient to get more gold so killed the goose, and when he cut it open, he found the innards of an ordinary goose, and …

Jackson winced, interrupted. "I wasn't suggesting killing, merely appropriately anesthetizing one and opening the skull for purposes of study."

Gathering her usual ability to confront, Elizabeth continued, "And how do you know that this can be done to an unknown species without causing severe bodily harm? Alon is not a fish. He is a sentient intelligent being. Suppose Martians discovered us, wanted to know how we communicated by making noises, and captured you to open your skull?"

Jackson's face turned beet red. "That's not the same!" He shuffled some papers. "Sorry, but I have another appointment."

Elizabeth had just burned to a black crisp her plan to convince Jackson. "I hope we can continue this conversation another time," she said politely as she stood, nodded, and left.

Shit! Stupid! Stupid! Stupid! she mentally chastised herself. The Golden Goose analogy had popped into her brain as a good one, but the farmer was a fool, so she realized she was intimating that Jackson was a fool. It's back to the drawing board. How she wished she could consult with Pierre. She could only hope Jackson would calm down and give her a chance later to pose her idea of attempting better communication with Alon.

She decided to walk back to the hotel to work off her feeling of failure. She should be practicing sending messages to other people as it seemed Alon implied that it was a matter of mind over matter. She laughed as she concentrated on the man walking ten feet in front of her in the same direction, "turn around, turn around, turn around" she tried to telemessage. Nope! Now she had to decide whether to let Jackson cool off and try again or launch a public campaign against capturing an Alonite ASAP. She didn't have much time left on her break. Her brain waltzed around the problem, as if she were playing musical chairs with the chairs being her options. The music stopped on the chair labeled abandoning the idea that she could reason with Jackson. That didn't mean she shouldn't continue communicating with Alon. Tick. It also meant she should start a campaign to stop Jackson before she returned to Yellow Submarine. Tock. She needed a drink.

By the time she reached her hotel, the walk and decisions made had calmed her. She no longer needed a drink and headed for the business room to begin to compose an op-ed to rally the public. First, she composed a note to Jackson thanking him for listening to her and apologizing if he thought she was intimating he was like the farmer who owned the golden goose. It couldn't hurt. She worked on the op-ed for two and a half hours before taking a break for a cocktail and a delicious meal.

Ideas had been percolating in her brain during her entire break with time out only to mourn her mother with her family. Her sisters were good judges, and she hadn't talked to either of them since her mother's funeral, which now seemed months ago. She couldn't contact Pierre, who was hiking in Cape Cod National Park. Until Elizabeth joined Yellow Submarine, she and her sisters had regular Zoom calls. She texted them on her burner cell to see when they were available, explaining she'd like their opinion on something as soon as possible, and emailed the op-ed from the business room in the hotel. She and her sisters were close in age but different from each other, which could prove advantageous. She planned to send it to appropriate columnists and invite them to take up the cause before trying to get her op-ed published. In fact, it would be ideal if they did not quote her.

She was able to set up the Zoom for the next morning. She'd also sleep on what she wrote, which inevitably led to a better piece be it an essay, a letter of complaint, or another important document.

Finally, she treated herself to a sumptuous dinner at the hotel and relaxed in its spa afterward. Both were heavenly therapeutic; her confidence in doing the right thing blossomed. Before bed, she made a list of potential columnists from the most widely distributed newspapers.

Chapter 70

(Elizabeth is on fire)

After breakfast, Elizabeth headed toward the business room hoping no one else would be there. She was lucky. The sisters popped on Zoom at about the same time. Elizabeth understood they would want to catch up on family affairs first.

Margaret piped up, "Elizabeth, you wouldn't believe how thrilled Pearl was with the uniform you wear on the submarine. She's the envy of her entire high school because it's authentic, not a knockoff."

Elizabeth laughed, thinking how mundane she and the entire team considered them, especially wearing them day after day after day. While designed to be unisex, for some reason only women and girls thought them fashionable.

Suzanne's sons were doing well. The sisters talked to Johnny less often after the death of their mother, but they said there was no news on that front except Johnny told them he was John to his friends. When they got down to discussing the op-ed, both sisters had editorial suggestions where they couldn't follow the content, but the important thing was that both shared her outrage at the idea of doing exploratory surgery on the brain of this new species because, of course, that's exactly the reaction she intended.

"Thanks, guys" Elizabeth said. "You were a great help."

Elizabeth spent the rest of the morning in the hotel's computer room tweaking her message depending on the columnist's newspaper. She made it clear she was appealing to a list by including the other columnists in a cc, and that she'd respect any newspaper's exclusive policy. No one would want to be accused of plagiarism even if each would make the points differently. She said that she'd accept the first to agree to write about the issue, and would submit as an op-ed if no one chose to broach the subject. So determined, she also reversed course on allowing her name to be used as that would provide greater credibility than if they wrote the source was a member of the Yellow Submarine ROE team. Jackson would guess the source anyway. She emphasized the fact that she was the one who interacted the most with Alon. Since the public was highly interested in this new species, it seemed natural for a columnist to be interested in writing about the topic. However, she inherited her mother's superstition that counting on something to happen invoked bad luck as in "Don't count your chickens before they hatch."

Elizabeth heard back from two of the younger columnists within two days. They both shared her indignation, agreed to share publishing on the same day in four days. That gave her a few days to relax and return to Yellow Submarine before publication. She visited the fabulous Metropolitan Art Museum, wending her way through unusual and provocative paintings like a cave woman checking out an abandoned cave. Afterward she strolled through the wooded parts of Central Park, feeling she had been exported back in time before humans had evolved.

She knew war would break out with Jackson.

Chapter 71

(Back home on Yellow Submarine)

Time flew like a hen escaping a rooster, and soon she was boarding the scaphe back to Yellow Submarine. While anticipating Jackson's wrath, she couldn't help feeling a perverse pleasure about going public. Her timing would appear sneaky, but she had had enough of New York City, and looked forward to returning to the fold. Yellow Submarine was her home. Further, she had a chance to resume meetings with Alon at the old location. One drawback is that she wouldn't get to see Pierre before returning, and she missed him. Maybe it was just as well given Rule # 9. It will be difficult back on the submarine. They'd have to settle for infrequent hugs, hugs that needed to be seen as friendliness rather than evidence of a relationship. She didn't need Jackson to have an excuse for firing her.

All aboard for home, she chatted idly with the pilot and aide as no one else was returning from leave. They seemed genuinely interested in how she liked her job, and life down there. When he realized she had experienced the bathyscaphe incident, he lapped up the details. Now, like her mother's funeral, it seemed long ago.

"I guess there's never a dull moment down there," the pilot said.

"Not many, but we do fall into routines. Exploring in the caterpillar and strolling on the upper deck are the only times I feel I'm in the ocean. Elsewhere it's like a dorm building. Actually, it's more like being in a science building with a huge aquarium although sometimes you feel you're the ones in the aquarium because the fish will swim close to the glass and peer in."

The pilot chuckled. I like your term for a bathyscaphe. "How's the food? Are we bringing you enough variety?"

"It's pretty good. There's not a lot of variety, but we have a guy who loves to cook, and he comes up with different ways of preparing things. Then we're getting more from the plants in the nursery. All in all, I'd have to say that we humans can look forward to survival via submarines."

The pilot seemed to sober. "So, you think we're all going to be living on one someday?"

"Maybe not all, but a substantial percentage if population growth continues even at the current slow pace. Of course, a smaller population would lessen the need for submarines. I remember when I was a kid, there was a TV show about people living under a giant round glass roof. I think it was called The Dome. It was based on a story by one of my favorite authors, Stephen King."

"I loved his books. He was prolific all the way up until last breath," the pilot said.

Elizabeth had been enjoying sharing with him and was surprised when they arrived. As soon as the door opened, she met Yusuf planted in the doorway waiting for her.

"Welcome, Elizabeth. We need you. No one else is interested in exploring. In fact, I order you to learn to pilot. It is dangerous to have only one pilot on board the bathyscaphe." Only Yusuf refused to call it the caterpillar.

She agreed, and she didn't understand her reluctance to pilot given the farm machinery she operated growing up. "You're right."

"Tyrone will teach you beginning tomorrow. I will see you at our dinner meeting tonight." Mission accomplished, he turned as if in a revolving door and sped away.

She had decided she didn't need to tell anyone about her last meeting with Alon. Sometimes, she didn't know why she did or did not do something. While she didn't believe in operating from one's gut as a rule, sometimes it was not a bad idea.

She retired to her closet of a room determined to rest. Her last days in NYC after getting her op-ed approved had been consumed with finding means to communicate with Alon. She had spent hours in a public library, Barnes & Noble, and Bookoff. She bought picture books of plants, animals, city infrastructure, and people working at various jobs along with a large laptop and downloaded movies that displayed human behavior.

She purchased Helen Keller's book *The Story of My Life* and two books on nonverbal communication. She considered how Helen could have learned abstract words a mystery.

Chapter 72

(Elizabeth learns shocking news)

Elizabeth didn't sleep well with wild and disturbing dreams, the classic ones of running from a monster but not moving forward. The monster was prehistoric, and then suddenly she was trapped in an elevator by herself. She screamed herself hoarse, then heard Jackson cackling maniacally, and finally woke in relief. It may be the only time she welcomed her cramped quarters cradling her.

Despite the day involving nothing unusual except for Tyrone giving her her first piloting lesson, Elizabeth felt on edge. She was waiting for Yusuf to call her into the situation room after an explosion of anger from Jackson being dumped on him. She sat at his table at dinner out of sympathy that he sat alone. She appeased his curiosity by describing what she had done and seen on her break. He expressed his condolences about her mother and for some reason the reality of her sorrow sunk lower. When tears welled in her eyes, Yusuf lowered his head as if not knowing what to say, then came up with "It's strange you sacrificed two days of your leave to explore our new territory."

Elizabeth just shrugged.

They finished their meal in silence as if Yusuf understood another person's need for peace. He wasn't a complete blowhard. He stood. "Well, it is time for me to see if Command Center has any new orders. I haven't had a chance all day to check with them.

Elizabeth gulped, but she should have known there was a reason for Yusuf's apparent lack of awareness of the editorials in opposition to Jackson's goal. She remained in the dining room as many people did at day's end.

She joined Haru and Heather sharing jokes. Everyone else had left. Heather told one about the wife of a prominent liberal German politician who shocked him when she said she opposed same sex marriage. Yes, the wife said, "I'm tired of the same old sex." Then they started making fun of Yusuf's arrogance but hushed when he returned to the dining room. Once he spotted Elizabeth, he walked over. Join me in the Situation Room. He turned and left before Elizabeth could close her mouth.

"Uh oh," Heather said. "It sounds as if Elizabeth has been naughty."

"As a matter of fact, at least in Jackson's mind, I'm responsible for two columns condemning surgically examining the brain of an Alonite." They were scheduled to be published yesterday."

"How will anyone know you're responsible?" Heather asked.

"I gave them permission to list me as a source."

Heather winced. Haru looked stricken.

"Jackson would have figured it out anyway. He knows I'm opposed, and I was in New York. One column was in the New York Times."

"Oh well, what can Jackson do to you?" Heather asked.

Elizabeth shrugged. "Sorry, but I think I'll go find out." She headed for the Situation Room.

She knew how Jackson would react but worried about Yusuf as she tapped on the door before she let herself in. "You've heard, I gather. I knew Jackson would be furious. I'm sorry you had to hear him let off steam. What did he say?"

Yusuf got to the point. "That at the next meeting of the Command Center he's going to recommend you be put on probation."

Elizabeth couldn't breathe for what felt like an entire lifetime. She had maintained her cool when her life was threatened in the damaged caterpillar, but at this news her bold nature flew away like a startled bird on a hot wire. After what seemed like enough time for water to freeze, she asked what basis Jackson planned to give to the committee.

"Obstruction of the advancement of science as forbidden in the bylaws of members of the UN Science Program," Yusuf said. "I am sorry, but I agree with him."

"So, what does probation mean?"

"He gave no details, and this is a first. However, I think if the rest of Command Center agrees, and you try to stop him from his goal regarding the Alonites in any way, you will be fired. This is serious, Elizabeth, and I know how much you love being part of the Yellow Submarine project."

Elizabeth's face turned white. "I hope you put in a good word for me."

"I told him that your position is an ethical one, but that I agreed that the advancement of science superseded your perceived violation of ethics in this case."

A speechless Elizabeth felt as if the blood had drained from her body. She had become a rag doll.

Jackson seemed pleased that I agreed with him and pointed out that scientists have used monkeys and rats to develop medicines and to learn more about how bodies work.

"That's not the same thing," Elizabeth said in righteous anger that Alon was not an animal. "We are dealing with intelligent beings not different from us. This is more like what was done to Blacks back when studying venereal disease cures. That's finally rightly recognized as immoral and beyond disgusting."

"Jackson said the columnist for the NYTimes said the same thing. Ironically, that comment got him particularly angry."

"That means it hit home. I'm sorry, but I need to go walk the deck to calm myself. Good thing we don't have a plank."

Yusuf said nothing for a moment. "Would you like a day off? I am willing to grant you one."

Elizabeth considered for a moment. "I'll let you know later."

Elizabeth strode so fast for the first lap that she passed several team and non-team members. Finally, she slowed down to a normal pace when she caught up with Haru who asked what Yusuf wanted.

"Jackson plans to recommend putting me on probation to Command Center.

"What!?

"Yes, Jackson blew up over the columns, and Yusuf agrees with him." I …

"I am glad you told me. Too bad Sanjay isn't back yet. He would have told Jackson that the rest of the ROE team agrees with you. Can you apologize for going public?"

"No, I can't because I'm not sorry." Elizabeth contemplated. She wasn't willing to give up the fight nor life on the submarine. The question was whether she had done enough. After a few moments, she continued. "I can say I won't keep fighting Jackson, but I believe there are better ways to learn about Alon's ability to telemessage than brain surgery. In fact, you know I'm working on means of direct communication with him. I've brought videos, pictures, etc. in case I see Alon again. Would you ask Yusuf if he'll transmit this to Jackson? I did try to reason with him before I went to reporters, but I blew it. I upset him when I reminded him of the story of the goose that laid the golden eggs."

"That's a good analogy, but not a good thing to say to him. Your better angels must have been sleeping."

"If I have better angels. I just blurted it out." Elizabeth managed a slight smile.

"Are you feeling a little better now?"

"Well, not OK but I am more calm." She sighed. "He didn't say how long the probation will last."

"We should check the bylaws."

Elizabeth sighed. "I'm going to do a couple more laps. Don't feel the need to go with me."

"I'll go check the bylaws for you."

Elizabeth did five more laps while plotting how she could work undercover and pondered over whether she hadn't set in motion all that needed to be done. She was dying to know the public reaction to the two articles and hoped Pierre might pick up some sense of it. He must be back in NYC. She hoped he'd seen the article.

Chapter 73

(Helen Keller's secret remains secret)

Within hours, Elizabeth recovered her stubborn self. She began reading Helen Keller's life story. To her dismay, there was nothing, absolutely nothing, about how she learned abstract words. She was impressed, however, at how her sense of smell and touch were exceptional — the only two senses on which she could rely. She couldn't imagine being able to lip-read by touching someone's lips or being taught to speak using touch. Helen could sense leaves waving in the wind and a difference in the air under full moons.

Obviously, she was a very bright woman. If you can't point at naughty or nice, with enough examples, small children can abstract to the quality. Helen must have learned the same way. You can't formally define the word *two* to a child by saying that it's the cardinality of a set that can be put in one-to-one correspondence with the set {one, two}, nor can you assume that a child who can recite one through ten can count. In fact, one of Elizabeth's little nephews once recited one through six while randomly pointing at his fingers to conclude he had six fingers on his hand. However, she heard her niece, Pearl, watching Sesame Street where a clown held up and announced four pie tins, four pie crusts, and four apple pies carried on a large tray. When the clown dropped the tray, Pearl said "four messes."

She was impressed with Helen's persistence and decided that might be the quality she needed to rely on. Persistence is a nicer word for stubbornness. Let it be enough for her to stop Jackson.

Chapter 74

(Command Center Meeting)

Jackson passed out the agenda, including a report indicating that Yellow Submarine had successfully returned to its original site with everything in place.

The three items on the agenda were:

#1 Applying to the National Science Foundation in the US and the Worldwide Science Foundation for funds to do deep dives to study one of the sea aliens

#2 Placing Elizabeth Gayer on probation

#3 Mitigating the damage done by two articles in the NYTimes and WaPo opposing learning more about the deep-sea creatures

Jackson asked for a motion to accept the report and received one with a second and unanimous approval.

"Onto the first item on the agenda: our budget does not support deep dives to search for the deep-sea creatures. The NSF and WSF are the natural agencies to fund large scientific projects."

"But they do not have that kind of money," several members chimed.

"No, but they are the entry to funds allocated by Congress and the UN," Jackson replied.

"That's presuming we've agreed to capture one of the aliens," the oldest member said.

"What? You don't think the finding that intelligent, sentient beings live under the sea floor who can telemessage isn't the most significant scientific find of the century?" Jackson said leaning over the head of the table.

"Of course, I do, but Yellow Submarine explorations have already discovered a good deal about their governance and way of life and are likely to uncover more. We haven't yet decided to launch a separate capture effort. If not, we don't need extra funds," said the oldest member of the committee. The other members attentively followed the disagreement.

Jackson took a deep breath. "Let me be clearer. Your chair moves that the Command Center write proposals to the NSF and WSF to provide funds for deep dives in submersibles that are able to investigate an alien brain."

"And how can that possibly be accomplished when both capsules are sealed tight to preserve the different pressures that humans and this other life are accustomed to?" said the oldest member. At his advanced age, he had heard it all and felt no social pressure not to tell it like it is.

"Indeed, it would be a major engineering problem, but that's why it will take considerable funding?"

"But what's the overarching idea? Somehow surround the alien submersible with a gigantic container, and then somehow robotically pierce the alien vessel with a tube through which a robot arm will anesthetize the alien and probe its body?" The other five members' alarmed faces swung back and forth between the quarreling men, both of whom had long commanded respect.

"Something like that, yes." Jackson squirmed.

"Do you realize how absurd that sounds?" The elder barked.

"Well, you proposed it." Jackson's face grew redder and redder by the second.

"The public already views scientists as hopeless eggheads. I urge a no vote." He turned to see who might follow his lead.

Jackson steamed but managed to keep the sound level down if not the tone of his voice. "Then, let me propose instead that we seek funds to reward the most feasible plan any company can devise."

The elder merely rolled his eyes.

"Any opposition to that idea?" Jackson said in a tone that dared dissent. "Hearing none, let's vote. All in favor? Three hands raised. Opposed? The remaining three members raised their hands. My vote is yes, breaking the tie. Now, onto an easier matter, your chair moves that Elizabeth Gayer be put on probation for her role in inspiring or otherwise motivating the two articles in opposition to the progress of science?" After an awkward silence, Jackson received a second from his closest ally. "Any discussion?"

"I found the arguments in both articles quite compelling," offered the chubbiest board member. Another Board member agreed.

Jackson replied, "Once more, imagine if humans could be altered to telemessage. It would be a leap for humankind beyond telephones, texting, and emailing. There would be no downtimes, which could be critical should our country ever be dragged into war."

The tallest board member asked incredulously, "Are these sea creatures able to telemessage over long distances?"

"I don't know. That's what we need to investigate," Jackson said.

"Do we know that particular individuals can be messaged and whether it is safe from interception by an enemy," asked the only woman on the board.

Jackson beamed "Again, that is why we need to know more about this ability. What if we were at war with these aliens?"

The elder smirked at his private thoughts. The board members remained silent while Jackson scanned their faces for further questions and reactions.

"Are we ready to vote on the probation? The discussion seems to have drifted from Ms. Gayer to the validity of the articles."

"Aren't they related?" the woman asked. "Have we ever had a vote on the issue of capture and crack open a head?"

Jackson bristled. "Yes, we just voted positively on proceeding to learn how they telemessage. I don't see any other way of accomplishing this goal."

"What does the ROE team say to this?" the woman continued. "In fact, I'd like to know if Ms. Gayer or linguists think there is another way."

"I heartily agree," spoke up the oldest member again.

"We could investigate other possibilities, but according to Ms. Gayer we don't know whether we'll ever encounter Alon or one of his fellow sea creatures again. We need to wait before we do anything. Further, Ms. Gayer has convinced me it is immoral to capture an Alonite," the woman continued. "Further, she suggests we continue to learn about them from the alien's mouth."

Three of the board members frowned but said nothing. The other three sat stone-faced.

"Are we ready to vote?" Jackson said barely keeping the edge of anger out of his voice. He took silence as agreement and called for a raise of hands. The vote was three/three again, and once again Jackson broke the tie, succeeding in going forward to fund exploration of means to study Alonites.

Jackson turned to the next item of putting Elizabeth on probation. He began with the argument that she had disrespected Command Center in not getting approval to go to the press.

"Would you have approved the letter?" the woman asked. Getting a sound no, she continued "It seems to me she was acting on her view of what science is about or rather not about, and she has a right to do so. It was clear she wasn't representing Command Center."

After some discussion, Jackson claimed "It's time to vote." After another 3/3 vote, he broke the tie and said he'd inform Yusuf.

The oldest member moved to postpone the 3rd item and this motion was unanimously approved.

After Jackson strode out of the meeting room, the rest sat for a moment. "I think we saw a side of Jackson we've never seen before," the woman said soberly. One of the two other No voters nodded his head, and the Yes voters were silent.

"I have a bad feeling about this," said the other No voter.

"I hope we don't regret this," said one of the Yes voters with apparent agreement written on the other two Yes voters' faces.

"And I hope you do," the woman said,

They all somberly left the room.

Chapter 75

(Elizabeth waits)

Elizabeth took the decision in stride. Sanjay had once told her that Command Center inevitably went along with Jackson. She consoled herself with the fact that there was nothing she could realistically accomplish now anyway. Other team members suggested she had nothing to worry about. Heather half-joked, "Just behave to please Yusuf, and you should be good to go."

Resigned, she became determined to keep busy, and she looked forward to Pierre's return but felt some trepidation.

Hans always prepared a special meal for those returning to the submarine. Elizabeth signed up to help with the meal, welcoming him and Pierre. The mixed green salad would be Pierre's first clue of how well the nursery was doing.

Pierre provided Elizabeth with relief when upon his arrival, he paid no special notice of her, greeting her the same as he did the others. They sat at separate tables. Alessandra sat at his table. She heard him say he'd had a long day and needed to retire early. Alessandra offered to walk with him.

Elizabeth sat with Heather and Hans, who reassured Elizabeth that she'd be reinstated if she kept a low profile. To her surprise, Pierre returned, saying he had left his briefcase under the table. He picked it up and said, "Before I retire, I have to check out the nursery."

"I'll go with you," Elizabeth said. "I want to see your face when you see how much the newest plants have grown." She raised her eyebrows at Hans and Heather in a gesture of inquiry about who might want to join them. Elizabeth contained her pleasure at shakes of heads. Heather added that she was taking the first lift up to her room.

Once inside the nursery, Pierre hugged Elizabeth. "I missed you. I am so sorry about your probation."

Elizabeth smiled, "Thank you. You have to know I missed you too." She pulled away from Pierre, reached into her uniform pocket for a slip of paper and put it in his hand. She had listed what she thought was appropriate behavior, prefacing it with an admittance of being paranoid because of her special need to toe the line. In sum, she allowed hugs, but nothing that could be interpreted as romantic feelings toward each other in case they were encountered. She didn't even want romantic talk.

Pierre blinked as if a mosquito had flown into his eye. He opened his mouth as if about to say something, then closed it again. "Thank you for showing me around New York City. I had a *wonderful* time." His eyes sparkled his feelings, a classic look of a rogue.

"I'm glad," Elizabeth said, her face revealing she understood he was not speaking of Central Park.

"Have you found peace with your mother's passing?" he asked, initiating a safe topic.

"I have, and it soothed me to share memories of Mom and see how my niece and nephews have grown." She paused and adopted a stiff upper lip. She was a good mother. Thanks for asking."

"Do you ever want children?" Pierre clasped his hands around her cheeks and turned her face toward himself.

"I wouldn't completely rule it out, but what's important to me now is contributing to the ability of the human race to survive. I don't want to bring anyone into an unlivable world. It's a decision I don't want to even think about yet."

Pierre nodded. "I understand."

Elizabeth tried to stifle her feelings, but the way Pierre locked eyes with her made her feel weak and as close to physical intimacy as one could get without touching. Then he stood next to her and softly patted her right butt cheek, saying, "These tomatoes look most appealing, simply delicious. In fact, every single thing here looks in decent shape," he said with a smirk on his face and twinkles in his eyes. "It appears I'm not needed."

"Oh, I think you are needed, very much so." She tilted her head back as if inviting a kiss. After locking eyes again for enough time for plants to grow, she said reluctantly, "You'd better get some rest. Travel can be exhausting." She shook herself back to reality and started for the door. "Oh, you should go out tomorrow with Tyrone. He's going on break soon, and with the nursery doing well, you can have Tyrone continue his schooling on piloting the caterpillar. Yusuf has ruled everyone who goes out has to learn how to pilot. That means me as well."

Pierre nodded while putting his arm around her shoulders. Elizabeth's entire body tingled at his touch. She felt grateful the French were known for being touchy-feely. They could claim any detected behavior to be the French way.

Chapter 76

(Things calm down for Elizabeth)

She and Pierre studied the caterpillar manual together in the library. She wondered why she had shied away from piloting. The library, while well-stocked wasn't a popular place to read, and the couple enjoyed the intimacy of being able to speak freely if not act so.

Elizabeth and Pierre also shared pleasure over the nursery's progress. Salad greens, tomatoes, green beans, peas, and other vegetables were thriving like five-year-olds released on a playground. Alessandra frequently dropped in, but she would leave when she saw Pierre was not alone.

Elizabeth and Tyrone had become closer after their near escape from death. She wanted to believe that sharing his heart-breaking experience was healing, not just hopeful thinking on her part. Life must be draining when you cannot trust the people around you.

On one of their explorations, Tyrone filled in Elizabeth on how the team reacted when they lost contact with the caterpillar. "You returned to your break so soon you never heard how upset people were. Heather was biting her nails, and Haru was beside himself. He even felt guilty for being involved in the choice of Yellow Submarine's new place. They talked about how they missed you when you were on leave." Elizabeth felt her heart melt when he said, "Their concern over losing both of us made me feel accepted in a way that I never have in a predominantly white group of people."

Elizabeth pumped her fists like a little boy smacking his first ball with a bat. "Finally, you feel like you're one of us because you are."

Tyrone responded with a grin, while Elizabeth wiped the wet streaks on her cheeks.

Chapter 77
(NSF Board meeting on The Alien Project)

The chair of the ten-member board of the National Science Foundation sat at the head of the oblong oak table and passed out the agenda, including the minutes of the last meeting. After a motion and vote to approve, the minutes were accepted. The chair continued, "We have only one item on today's agenda; the proposal may be controversial. Discussion is in order to recommend funding to capture an Alonite in order to study its brain and determine its ability to telemessage."

The woman to the chair's immediate left raised her hand. After the chair's nod, she said, "I'd like to know if everyone has read the article in the NYTimes or the Washington Post about the opposition to the project by one of the ROE team from Yellow Submarine." After a unanimous nod of heads along with a who-hasn't chuckle, she continued. "I, for one, was completely swayed by the argument that the idea lacks humanity and could well backfire. According to Elizabeth Gayer, Alonites are equal to us in intelligence. That they've created something similar to our bathyscaphe to conduct research confirms this. Would we do brain surgery on the world's best leaders to see if we can breed more good leaders?"

Most of the committee members chuckled, but two remained stone-faced.

The chair asked for further comments in a 'let's move on' tone of voice. One of the stone-faced members spoke up. "Examining the brains of human beings is different from examining some creatures that live below the ocean floor. Imagine if we humans could telemessage. It would do away with learning other languages, as I understand that those who experienced telemessaging reported it came through in their native language. That's nothing short of amazing."

The oldest member of the committee spoke up. "Remember when AI advanced to the point ten years ago of being dangerous? Two workers accidentally left a computer running on AI overnight, and it had taken over to the point that the only way to stop it was to unplug it. Having the capacity to sound like particular people is scary. The shysters claiming someone's grandchild needed money to get out of jail were even more successful using this power. Fake orders from the President himself and CEOs wreaked major havoc until Congress finally passed laws that severely limited AI's use. Fortunately, both parties were in rare agreement, or we'd probably have started WW III."

The second stone-faced member spoke up. "You just proved that Congress could keep telemessaging from being harmful."

"Could you explain how?" another woman said. "In fact, AI is still harmful since those up to no good don't pay attention to laws."

Stone-faced two stuttered, "Um, there'd be some way. It would depend on how it worked."

"You're ignoring the decency of acting toward others the way you would have others act toward you. I doubt you'd consent to such surgery to satisfy someone about how your brain works," said the only bald-headed member."

When all but the two who spoke in favor nodded heads, the chair asked if the committee was ready to vote. Hearing no objections, he called for a straw vote, which resulted in eight opposed and two in favor. "Would those opposed like to make further comments before we take a final vote?"

The two looked at each other and shrugged. "I can only say that I'm representing the opinion of those I've consulted with," said one. "As am I," said the other.

"Any other comments?" the chair continued. "Hearing none, let's take a final vote, again by a show of hands." The vote remained the same. "I will not pass on the request to Congress, but I will inform the relevant parties that we have considered and rejected the proposal."

After the meeting adjourned, the two women agreed to have lunch together. On their walk to a nearby restaurant, they laughed about the possibilities of the organizations those in favor consulted. "Yes, I wonder what company's CEOs they consulted?" the older woman said.

"Oh, maybe a company that might build large underwater ships or bathyscaphes to locate the Alonites," the younger one said in a voice faking a wild guess.

"Yes, progress doesn't benefit everyone equally. Technology could have shortened the workweek, but companies hired fewer employees instead, and their profits soared," said the older woman.

"Maybe this time, the execs' feelings will be *sore* with a different spelling of soar," the younger said.

They both laughed and headed for a seafood restaurant where they vowed not to eat their famous octopus.

Chapter 78

(Jackson doesn't give up)

A similar meeting with the same outcome took place with the board of the World Science Foundation, but Jackson was not deterred. He visited Senators and members of the House of Representatives on appropriation committees and lobbied them to propose a bill jointly. He asked for it to be labeled The NEW AI. He kept Yusuf up to date, thinking the ROE team would approve of his persistence. Instead, they suspected Jackson's real motive for trying to find out the secret of telemessaging was to harness it for profit. His rich family could monetize telemessaging and become so rich they'd control the World.

Elizabeth and her colleagues celebrated when they learned that Jackson's measure did not get enough support to survive the floor votes in both the House and the Senate. This is despite Jackson's family donating money royally to every member of the appropriation committees of both houses.

Jackson, however, continued to work on the cause. In fact, now he had reason to ask for private companies to donate, and his family knew just which companies they'd supported that would be more than interested, particularly if promised part of the profit from companies managing telemessaging.

Chapter 79

(Sanjay works on Command Center)

Sanjay slept ten hours at night and napped in the mid-afternoon at the hotel near Command Center's headquarters for the first week. His energy quickly returned, and he hoped his case would add to the evidence that sleep is the ultimate cure-all. Once his need for sleep returned to normal, he became bored and found himself visiting Command Center members other than Jackson and lunching with them. He voiced his opinion about the inhumanity of trying to examine an Alonite's brain. He hoped as the ROE team leader, his opinion would carry weight. He wisely requested that those he contacted not quote him personally about the inhumanity, rather indicate it was the overwhelming opinion of the ROE team.

When he learned of Elizabeth's probation, he was dumbfounded even after reading the article motivated by Elizabeth. By the time of the meeting when reinstating Elizabeth was to be considered, he had painted an excellent picture of her dedication and value to the ROE team.

When he learned of Yusuf's support for Jackson's plan, he became even more determined to return to Yellow Submarine as soon as possible. The physician who served Command Center agreed he was ready, and Sanjay convinced him to convey that his well-being depended on returning to his leadership of the ROE team.

Chapter 80

(Decision on Elizabeth's reinstatement and Sanjay's return)

Director Jackson called the meeting to order. The routine business list and the last meeting minutes were quickly approved.

"Then on to the first item of business. Sanjay is eager to return to Yellow Submarine. He has been a good leader, and his doctor confirms that he can return to work. In fact, he argued, it was in Sanjay's best interest. Any motion to his return to the leadership of ROE?"

The motion was made, seconded, and swiftly approved.

Jackson turned to the controversial item. "This is the last day of Elizabeth Gayer's probation. We need to decide whether she should be dismissed, continue on probation, or be relieved of her position on Yellow Submarine."

"I've reviewed our bylaws," the only woman on the committee said, and they say unequivocally that continuing on probation is not an option."

Jackson frowned. "Then, I am arguing that she be dismissed. She has done nothing to stop the ongoing discussion in the press."

"Has she contributed any further articles on the subject of Alon?" The oldest member inquired.

"Perhaps not," Jackson said, "but columnists frequently quote her original article."

"And how is that her responsibility now?" The woman asked.

"She is a scientist," Jackson practically yelled.

"I call for the vote," the elder said.

"Agreed," Jackson said, eager to quell the argument.

There were two votes in favor of reinstatement and four refusals to vote.

Jackson's face radiated befuddlement, if not anger.

"In this case," the woman said, "the vote favors reinstatement. Again, this is spelled out in the bylaws."

"Have it your way," Jackson said. "I always follow the rules even when I think they're wrong." It was apparent he had to struggle to maintain his reputation as a respected leader. "Let's move on to the next item."

The rest of the business proceeded with no further animosity.

Chapter 81

(Elizabeth makes progress with Alon)

Yusuf had informed Elizabeth that her reinstatement was on the agenda of the next meeting of Command Center. While she understood that the amount of public discourse over concern about Alon was out of her hands, she feared it might work against her. She deemed working on improving communication between herself and Alon as her best recourse for staying calm. Then she snickered as she thought she had less chance of being able to telemessage than swimming across the Indian Ocean. All of her life, she had been decisive, rarely ruminating over how to handle situations. Now, she felt like a person in the cartoons with a little red devil on one shoulder and a white-winged angel on the other; only it was worse because instead, there were two tiny, identical Elizabeths in the color of her uniform, debating whether she had a chance of substantive communication —no way of knowing which was the wiser.

She spent every waking moment thinking, even having dreams about her telemessaging Alon and them discussing their favorite fish to eat or which fish was the ugliest. When awake, she focused on visual communication since her efforts to send brain waves got her nowhere, not with her colleagues nor with Alon. She used a magnetic whiteboard to display lower-case letters and integers 1 through 9. She began with a photo of a dog next to a display of the spelling, D O G followed by C A T, and a few more animals, even mimicking their behavior. Alon had laughed at her antics, and Elizabeth seized the opportunity to pull down L, A, U, G, H, pointed at him and then feigned laughter. He telemessaged to verify that she showed him symbols for things in her world. She viewed his understanding of her intent as major progress. Nouns were easier to communicate because of pictures and the ability to point. Actions were a different kind of fish. The numbers were easy as she counted on her fingers and pointed at the symbols. She saw that Alon had only four fingers, so she wondered if it was possible they had a base numbering system with eight as the base.

By the time she learned her probation had been lifted, Alon recognized twenty-three nouns, two pronouns, two verbs, two states of mind, and the symbols for Elizabeth, Tyrone, and Pierre. Not much, yet Elizabeth's hope of success increased. She held a large laptop to the caterpillar's window and showed movies of people walking, driving cars, going up steps and ladders, picking apples and berries, and eating them. Alon soon understood the few simple sentences one could convey using the words he knew. She could attest to his intelligence and doubted she could learn his language as quickly.

Then, something seemed to click, and Alon began to telemessage questions with yes or no answers. "Are there many different trees? Different food is on plants? Where is sky? Cars can't go up?" Elizabeth didn't think airplanes were important enough to try to explain, but then showed pictures.

Alon asked "You need special cars with attachments?"

Elizabeth was beside herself with excitement about what she regarded as a major breakthrough.

Alon still encouraged her to try to telemessage to no avail. After several unsuccessful attempts, she shook her head whenever he asked her to try. She shared successes with Tyrone, saying, "Alon takes to learning to read English like a fish to water." When Tyrone groaned, Elizabeth fake pouted. "I thought that was clever," she said. They agreed not to share Alon's progress just yet. Elizabeth didn't know why exactly, but probably the little Elizabeth on her left shoulder had taken charge. Tyrone rolled his eyes at her when she told him about her whimsical tiny Elizabeths, but he began to label her ideas accordingly, namely E-gut left and E-logic right.

Elizabeth's involvement with her project helped distract her from her aching feelings for Pierre and maintain the appearance that their relationship was platonic. While they enjoyed a significant amount of time alone in the nursery and library, it was genuinely work-related.

The scare in the caterpillar had lessened Tyrone's interest in exploring, and Pierre's satisfaction that Sydney had the nursery under control he could pilot regularly. She found Pierre helpful in suggesting ways of demonstrating the meanings of English words. Fixed in their seats, they could only express their feelings freely in the caterpillar. Further, piloting required attention. Just being alone together created a satisfactory sense of intimacy. Love is not a simple matter of sexual attraction.

Only in her dreams did she experience romantic pleasure with Pierre. Occasionally, she would express her thoughts at meetings about Rule # 9 but kept it abstract, merely suggesting that prohibiting married couples and partners from serving on Yellow Submarine didn't make sense. Of course, whenever she said anything, Yusuf would launch into all the reasons it did make sense, primarily that they all absolutely needed to work well together, and marital strife would obviously interfere. She refrained from saying her parents loved each other and worked well together most of the time. With Yusuf, the less you respond, the fewer words of pontification you'd have to endure.

Alvin Chapter 82

(Alvin reminisces)

Working on learning the alien's written language thrilled Alvin. He always loved learning new things. He assumed these aliens made sounds to communicate directly, as did his species, but hearing between different scaphes was impossible, so he was excited about aliens making marks to communicate. He had learned there were twenty-six marks, and her name's letters were ELIZABETH.

Alvin realized he missed his friends associated with Professor Thurgood. Friends with common goals connect in ways that tetrapeds thrown randomly together do not. Sharing a common enemy would bring tetrapeds even closer. Individually, every tetraped likes to think their existence makes a positive difference to society. All but a few of Alvin's like-minded friends had lived in Abyssal below Pelagia's floor since the government threatened them. He made a mental note to visit Monk — a local cop in Volcano City — and his partner, Angel. They were his only cohorts in thinking. He didn't live far from them, and he wondered why he hadn't maintained contact. He didn't even know if Monk still visited Abyssal and the rest of the professor's group.

The three enjoyed a meal of mashed blobfish together, and Alvin learned they had remained in contact with the old group. Monk and Lawson had been partners on

Volcano City's police force, and both fought for a cleaner environment. They were among the few in the professor's group who, along with himself, had not been discovered and identified as traitors.

He was pleased to learn Monk and Angel had scrimped and saved to purchase a filter for their cave, and that Angel was expecting. Monk's pride tickled him. Someday, *maybe*, Alvin thought. If it weren't for his discovery of the aliens that lived above them, he'd be looking to do something else, maybe even joining the Blues in Abyssal.

Chapter 83

(Elizabeth continues to work on communication with Alon)

When Alon began to ask good questions, Elizabeth began skipping on her walks around the upper deck. He telemessaged that he couldn't imagine what it was like not being able to rise above their ground. It seemed so limiting to him. Elizabeth shook her head in amusement, and he presciently observed that he guessed that one doesn't miss what one never had. By sheer coincidence, she had chosen the day's lesson to be about the sun, moon, and stars. The names weren't as important as helping Alon understand the nature of the human environment. She used photos of people looking up and pointing up at a heavenly object, the sun by day, and stars and moon by night. She lit a match to demonstrate fire and tried to connect sun and fire. Alon shook his head and frowned. Elizabeth gave up and understood for the first time the amazing nature of the sun. Further, how could he understand fire? His kind could see in the dark, the same dark that may have given birth to the many bioluminescent species.

Pierre had become an excellent pilot especially with his ability to detect potential strong currents before they were caught in one. He suggested teaching Alon the word KISS. Alon conveyed that his kind did something similar, but they did not laugh afterward. This caused more laughter puzzling Alon.

Elizabeth's enthusiastic efforts to teach Alon the written English language amused Pierre, but he joined her full force, providing ideas. With each effort to display human life, Elizabeth added accompanying appropriate adjectives like fast and slow. The big challenge, however, was how to ask Alon about his ability to telemessage. Especially curious was his claim that not all of his species could telemessage when he had suggested that she could learn. He had asked her to keep trying. Elizabeth realized that neither of them could have any way of knowing if humans could be taught.

Pierre suggested showing Alon a picture of the human brain. "Great idea," Elizabeth said. She had become creative with her hand motions as she worked with Alon. She found and held up pictures of the front, sides, and back of the brain, while matching them with her palm on the corresponding side of her head. He nodded knowingly, and she confirmed two conjectures. Alonites had similar-looking brain matter, and they had some means of photography since pictures didn't puzzle him.

Alon described a component of his species, one that lived in isolation from the rest and was the most proficient in telemessaging. They began teaching the young as soon as they started school. He had learned from them, but he had his brain kickstarted with a machine before he began his efforts. He admitted he knew nothing about the machine, and he shouldn't have considered the possibility Elizabeth could achieve the same goal without the procedure that connected his head to a machine.

Elizabeth couldn't believe the information she had just uncovered. She presumed that if a human child never learned language and lacked the ability to speak, it would be difficult to learn as an adult. If Helen Keller's deafness were cured, would it be easy for her to learn to talk? If not, this puts an entirely new perspective on Jackson's goal. In her mind, his vision of humans telemessaging just soared up and up away like a hot-air balloon.

Alon looked chagrined. He telemessaged "I know nothing about this machine. I think if you not learn when young, you need this. Sorry I say you try."

Pierre tapped Elizabeth's arm, signaling that they needed to turn back. Elizabeth nodded and slowly waved at Alon while embedded in her thoughts.

She sing-songed "Hallelujah, Deacon Jones" on the way back to Yellow Submarine to Pierre's amusement. She told him that her mother said this when something went her way. When she asked what it meant, her mom hunched her shoulders and said it was something her grandfather used to say.

"I guess you're pleased with today's progress," Pierre said.

"Am I ever…" She stared off into space. "It seems that if Alonites don't start to telemessage at an early age, they need special intervention to gain the ability. I guess it's like learning to read. It's easiest if you start young. I'm excited. If this information doesn't stop Jackson, nothing will."

"Yeah, but are you sure Jackson is only interested in Alon's telemessaging?" Pierre said as he turned the caterpillar back toward the submarine.

Soberly, Elizabeth pondered the question. "No, I'm not certain. He'd probably think putting Alon in an aquarium would be a wonderful way to teach science to kids."

They returned to the submarine to learn that Sanjay would be returning to take over the leadership with the next delivery. "Hallelujah, Deacon Jones," Pierre said, and Elizabeth couldn't stop laughing. She high-fived him.

Chapter 84

(Alon is making progress)

Alon's doubts about Elizabeth's ability to telemessage inspired her to work harder at helping him with the English written language. His progress was slow but steady. He now recognizes over five hundred written words.

Even when Tyrone occasionally piloted, Pierre joined them. He could assist both with piloting and Alon's education. Alon's progress surprised Tyrone. "You know," he said, "I'm not sure Jackson would be satisfied with answers to every question Command Center might have. We should be teaching him words to warn him if Jackson ever succeeded in manufacturing his wonder bathyscaphe."

Pierre agreed heartily, and Elizabeth shook her head. "I can't believe I didn't think of that."

"You were too focused on answering Jackson's questions," Pierre said.

"Wow, look at the size of that blobfish over to our right!" Tyrone pointed.

Their eyes could not leave the ugly fish until it was out of sight.

"That monstrosity almost made me forget my idea about marking our caterpillar and making sure Alon knows only our caterpillar has that mark," Elizabeth said. "Then, we must communicate that if he sees something without it, he should get away as fast as possible."

The three pumped fists. "I'm sure Sanjay won't have a problem with this," Elizabeth said. "All we need is a brightly colored slash below our windows."

"We'd have to have a way to remove it in case Jackson or a designate ever came down here," Tyrone said.

"Right," Elizabeth shook her fist. "Sanjay should be able to get a picture of Jackson that we can use to indicate Alon's enemy."

"Let's hope he never gets close enough to see him, or it may be too late," Pierre said.

"Yeah," Tyrone nodded.

"Duh," Elizabeth replied, laughing at herself. "However, he may remember having met Jackson, so his picture will help him understand where the threat is coming from."

This time, the three plotters high-fived each other.

Chapter 85

(Elizabeth solicits Sanjay's help)

Elizabeth asked Sanjay when he would have time to talk. He asked for fifteen minutes, and then they met in the situation room."

She went for a short stroll on the upper deck. It might be her imagination, but she thought she saw fewer angler fish. They must have lost their curiosity about human beings.

Sanjay welcomed her and shut the door. "How are you doing?"

"The question is, how are *you* doing?"

Sanjay laughed and gave a thumbs-up.

Elizabeth thumbed back. "I've been teaching Alon how to read English. I hold whiteboards with a word against my window and then demo its meaning. He's a quick learner. I'm hoping you can help convince Jackson that soon I may know something significant about how Alonite's ability to telemessage. It needs to come from you. I'm not on Jackson's favorite list."

"I'll do my best, but you know how determined Jackson is."

"I have another request. Can you get me a picture of Jackson and make a sign with a big red swoop to attach to the caterpillar? I want a way to warn Alon to stay away from scaphes other than ours."

Sanjay placed his hand over his mouth in contemplation. "The picture of Jackson is easy to get. I assume you want it the size of a window on the caterpillar. I'm just wondering about the sign if Jackson comes down here. Let me think about the best way. Something removable, or use yellow paint that dries quickly to cover over the sign? I'll figure something out. I won't take long."

"Awesome!" Elizabeth's face radiated with joy.

"So, tell me more about Alon learning written English."

"I can't believe how quick he is. If he is typical, then their society is more intelligent than ours. He only has trouble with things he can't fathom because his physical surroundings differ so much from ours. Fire and other planets are beyond his imagination."

"That sounds promising," Sanjay said. "Good luck."

Elizabeth left the situation room floating on air.

Chapter 86

(Elizabeth warns Alon)

Elizabeth felt compelled to inform Alon about the warning signs immediately. Tyrone and Pierre chided her as being obsessive, especially when rushing was unnecessary. They were right. Embedded in her DNA was the compulsion to complete any project in short order, like the need to scratch itches on her body. She understood the downside of this character trait and had sought ways to tone it down all of her life. She'd go for a hike at top speed, work on a puzzle, or some other distraction, during which she often experienced an AH HA moment that solved a problem bothering her. Having to wait a few days for Sanjay to obtain a picture of Jackson, she visited the library with a 1000-piece puzzle in the corner, then waltzed around the top deck multiple times.

Alessandra, the team's best artist, questioned Pierre daily about his explorations. She volunteered to create the sign. Although it needed only to be highly visible through the deep water, she went further and created an eye-catching artistic design using several swoops of different bright colors. She gave it to Pierre to pass on to Elizabeth.

While waiting for the Jackson photo, Elizabeth taught Alon '*go away*' using a doctored photo of a fish fleeing a shark.

Once Sanjay obtained a photo of Jackson, Elizabeth set out to instruct Alon on the most important communication she'd ever attempt. When Alon appeared, her anticipation lifted as if a puppeteer controlled her.

"I see you happy meet me," Alon messaged his canonical greeting.

Elizabeth nodded. Despite the nature of her message, she felt like her child self when she brought home a perfect score on a math test to show her parents. Taking a deep breath, she held up the photo of Jackson titled GO AWAY.

Alon's eyes grew large. He had no eyebrows to wrinkle or frown, but his eyeballs circled in their sockets for some time. Then he nodded soberly, causing Elizabeth and Pierre to hug. Tyrone cheered.

On the way back to Yellow Submarine, she realized that she should have conveyed that it would take a long time before Jackson could attempt his dire plan. Without that understanding, she might have seen the last of Alon. On the other hand, Jackson could try something rash without all of his dominoes in a row.

Nothing could be done but turn to her next goal of finding a way to inform Jackson that telemessaging required special equipment if you weren't born with the ability. If only Jackson could be permanently put off. Tyrone was due soon for a break, and Jackson might accept his arguments better than hers. Sexism may have gone underground, but it could be as dangerous as termites under your house destroying your wood supports. She suspected counting on Tyrone to succeed was as plausible as a bald man growing feathers, but the remote possibility was all she needed to relax for the time being.

Alvin Chapter 87
(Alvin's dilemma)

Alvin learned that the symbols ALON represented himself. Her name used the symbols E L IZ A B E T H. He told his current pilot, Haggard, that the being he was telemessaging was teaching him her language. Haggard understood and agreed to keep the amazing discovery to himself. Of course, given Fitz's reaction, Alon did not share that one of these beings wanted to capture one of their own. That caused Fitz to put his country first, as flawed as it is. He still felt guilt for ruining Fitz's life, but he had no choice.

Alvin did share with Haggard what he had learned about these aliens, and Haggard's scientific nature anticipated learning as much as possible about this spectacular discovery.

Haggard's eagerness to learn about another culture reminded Alvin of his fascination with the advanced society created by the Blues. Despite the quirkiness of the Blues leaders — Aylus and his brother, Skyler — Alvin decided to seek their sage advice. Further, they should be warned that the country founded by the Blues could also be vulnerable.

Alvin Chapter 88

(Alvin visits the leaders of the Blues)

Alvin set out for Abyssal to consult with Aylus and Skyler using the Gabbro Caves entrance. He was of two minds. One of his minds advised "no further communication with Elizabeth's kind, pretend she's a member of a school of sharks. The other mind said, work with her to learn what her kind is planning so his country can be prepared.

When he arrived at the domed offices of the two Abyssal leaders, a secretary expressed surprise. "I haven't seen you in forever. Are you here to see Aylus?"

"Yes, is there any chance of seeing him soon? It's urgent."

"Let me check." She turned her face from him and telemessaged the request privately. She smiled. "You're in luck. He'll be out in a few myks."

Alvin relaxed, and in so doing, realized how stressful he had found keeping his communication with the Blues from everyone in Pelagia. Of course, he wouldn't let Abyssal's leaders know he had broken his promise to Fitz. In his not-so-long-ago youth, he was the ultimate party kom, known to imbibe more than he should. As an aquanaut, the kems idolized him. He wasn't proud of his life before joining the underground environmental movement, but he smiled at the remembrance of his youthful shenanigans. At the same time, he felt a mental shake in his head at his immaturity.

He tore himself from his reminiscing and began to think something had happened to detain Aylus when Skyler appeared from the hallway. "Sorry, it took me so much time to get here. Aylus informed me you were here on a matter of some urgency." Alvin nodded and followed Skyler, who knocked briefly before entering. Aylus nodded at a chair while Skyler joined him behind his expansive desk. Aylus was the official leader of Abyssal, but on important matters, he always included his brother.

Without waiting to be asked, Alvin summarized his interactions with Elizabeth beginning with the first encounter. The brothers were so shocked, they nearly swam off their chairs. He had forgotten what a significant discovery she was, and was pummeled with questions before he could finally tell them about her recent warning. "If I understood this alien correctly, her kind want to examine one of our brains."

This warning did not surprise the brothers, and they grilled Alvin further until they were certain he had not given away their existence.

Once satisfied he had conveyed everything relevant, they allowed Alvin to explain his dilemma about how to proceed. He argued that on one hand, if Pelagia were to curb their explorations above their ceiling, the only evidence of the existence of Pelagia would be the discovery of the gigantic fan circulating cleaner water, and the chance of that happening was minimal since they could not travel that deep. Thus, with no further attempts to communicate, over time, these aliens would abandon searching for them. On the other hand, if communication between himself and Elizabeth improved, he could keep track of what was going on, allowing both Pelagia and Abyssal to be prepared. "I am certain that if these beings decide to investigate us it will take some time for them to invent stronger aquascaphes."

Neither brother spoke for some time, both frowning and lifting their heads as if to watch the fish swimming next to the dome over their heads. Then Alvin realized they were privately discussing via telemessaging what they'd just learned. Finally, Aylus said, "And you don't have an opinion on which of the scenarios you think most sensible."

"No. That's why I'm here. I trust your wisdom."

"How certain are you that there's time?" Aylus asked.

"Quite certain because as I said, they don't have aquascaphes that can withstand our pressure."

The brothers went into their private telemessaging again. This time, Skyler turned to Alvin. "We are thinking that it's not either/or. Suppose we find a way for Elizabeth to communicate what she can to inform us about their capabilities."

"Yes," Aylus said. "The more we know about them, the safer we will be, not only regarding their current plans but any that might develop. Further, at any point that it seems they are about to take action to capture one of us, you can disappear, and we can carefully guard our realms."

"That sounds wise to me," Alvin said.

"Then we agree?" Aylus said, and Alvin nodded.

"This is what I wanted. Now, I need advice on how to get Elizabeth to be able to answer our questions. We've made a good start."

"Do we understand she's been giving you lessons on her language by showing you pictures and corresponding symbols?" Aylus asked.

When Alvin nodded, Skyler stepped in. "Do you think you are in a position where you can take the lead? Ask her about symbols for words we need to use to inquire about what these aliens are planning and any progress they may have made. The foundation for you to learn her language has been established, so all that is needed is for you to proceed."

Alvin nodded again. "I've begun to do precisely that. I knew I was right about consulting with you. May I ask that *you* construct a list of appropriate words?"

Both brothers smiled. "Of course, you can ask." With a twinkle in Skyler's eye, he said, "We will think about it." Aylus nudged him. "Do not mind my mischievous brother. We will manage it; we will assign the task to our most brilliant scientists and linguists."

Alvin's exasperated sigh made the brothers both laugh.

Alvin felt like groaning. He had forgotten the brothers' penchant to be over-the-top literal, but he warmed to their next suggestion to check out the menu this eeken in the gorgeous dining room under the clear dome. He needed to celebrate. Abyssalian fish were free of contamination, exquisitely prepared, and made more delicious by the elegance of the dining room under the clear dome.

The brothers offered Alon a room in the government building, suggesting he could stay a few deks, visit old friends, and wait for their list of necessary words.

Alvin loved the idea. He had time before dinner to head for the city square that housed his former allies. He discovered Landley and Lawson had named their male offspring after him. The small kom's eyes widened when he learned that Alvin was an aquanaut and flooded him with questions. He loved it, and for the first time in his life, he considered what it would be like to be a father. He knew such thoughts needed to be postponed as battles were to be won.

Chapter 89

(Yusuf supports Jackson)

Because of Yusuf, Jackson's intention to capture one of Alon's species cast a pall on the team's comradeship. Elizabeth's private nickname, You Suffer, fit better than ever. Even when Jackson's idea wasn't on the agenda of the team meetings, You Suffer would bring it up as something the team should brainstorm about to assist the project. His argument was so repetitive that the team sing-songed it behind his back, waving their shoulders. "I, for one, understand that science is vastly more important than the life of one being. Yada yada yada!" A seething Elizabeth couldn't safely disagree at meetings. Meeting minutes were in the public record and no doubt scoured by Jackson for her name. She'd be devastated to be removed from the Yellow Submarine project.

Fortunately, Pierre and Tyrone frequently challenged Yusuf if no one else spoke up. Elizabeth knew that the majority agreed with her, but few liked to challenge old You Suffer lest they become the target of his viper tongue. Tyrone especially tickled Elizabeth because the more he spoke, the less reticent he grew.

Not having encountered Alon since she informed him of Jackson's objectives, she concluded she'd never see him again. Over a week had passed. Her feelings were mixed.

She both wanted him safe but rued the loss of their unusual friendship.

On a routine exploration with Pierre, Elizabeth shrieked so loud that she frightened Pierre until he saw Alon's scaphe appearing in the distance close to their usual meeting coordinates.

"You almost gave me a heart attack," Pierre joked.

Drawing close, Alon telemessaged, "I see again you happy meet me."

Elizabeth nodded her head. She told Pierre to dig up the whiteboard with the alphabet along the top.

"I want help you talk at me," Alon telemessaged. He smiled at her when her eyes grew wide. "I see you like."

She nodded.

"I telemessage a concept, and you show me word for it, like before, but I have a new list."

A beaming Elizabeth bobbed her head.

At the one-hour session, Elizabeth recognized the words as ones needed to warn him.

Elizabeth wondered whether Alon had a photographic memory. Adding the newly learned words, she had enough for a small dictionary, which she would include in her reports to Command Center as well as what she had learned about telemessaging, namely 1) not all Alonites can telemessage and 2) Alon's ability had to be kick-started by some machine. She had compiled a dictionary that enabled answers to the manure load of questions about telemessaging she anticipated Jackson might have.

Elizabeth met privately with Sanjay about her intentions, which included ways to curry Jackson's favor or at least to get her off his shit list. Sanjay told her he had been reporting Elizabeth's model behavior at every opportune moment.

Time was on Alon's side, given how difficult it would be for Jackson to find a sufficient number of willing donors and hire designers and manufacturers to develop a workable vessel. With favorable reports from Sanjay, she felt mildly optimistic she could regain Jackson's respect.

Keeping busy with her project made it easier to interact with Pierre in a platonic way. Their work schedules corresponded both with exploring and nursery duty. Their smiles, hand-blown kisses, and hand squeezes became sufficient to maintain how much they cared for each other. Just the way Pierre's eyes engaged her made her body tingle in all kinds of places. More importantly, their ability to work together established a more important bond than mere physical attraction.

The newly added fish traps to the back of the caterpillar and the bottom of Yellow Submarine near the openings for deliveries and caterpillars enhanced the team's diet. They had Alon to thank for the idea, and Sanjay assured Elizabeth that he had ensured Jackson knew it.

While all seemed well with Pierre, they had no commitment. Elizabeth found herself jealous when gorgeous Alessandra frequently popped into the nursery, offering to help. It was clear she hoped to find Pierre alone. Further, she couldn't tell if Pierre enjoyed her attention.

Chapter 90

(Congressional House considers funds for Jackson)

Intensely controversial HR 4922 opened for discussion on the floor of the House. The bill authorized 3.4 billion dollars to be added to the NSF budget to enable the invention and manufacture of a deep-sea vehicle that could travel a half mile deeper than had ever been ventured, at least not successfully. In addition, the vehicle would contain a special compartment to accommodate deep sea creatures at their normal pressure. The special compartment would contain yet-to-be-designed medical equipment built into its shell with the ability to conduct medical exams, particularly X-rays, ultrasounds, CT scans, PET scans, and most importantly, MRIs of brains. The current House consisted of 209 each of Democrats and Republicans with seventeen Independents. Support split precisely along party lines, with all Republicans in favor and all Democrats opposed.

The unique proposal had provided news columnists with fodder for months. Controversy always made popular news, and the media pandered to the public eye. The issue divided the country. The PBS Newshour reporter assigned to Congress, Aruna Kuman, summarized the arguments to date. She understood Congress better than 90% of her fellow reporters roving the halls of the Capitol building.

Her report on HR 4922 boiled down the arguments on the floor. "Members of the two parties — not for the first time — are in complete opposition. Thus, undecided Independents will determine the outcome. What's unusual is that the Republicans are in full support while all Democrats are opposed. Democrats are known to favor scientific progress, while Republicans inevitably oppose costly projects. Democrats argue that not only is the project next to impossible and inordinately expensive, but it is also inhumane. They accuse Republicans of being bought by the industries likely to profit. Republicans counter by extolling scientific progress and the importance of advancing understanding of our world. She described with amusement the reversal of typical positions of the two parties.

As the debate continued in Congress for several days, both pro and con guests were interviewed on news shows. Command Center Jackson watched them all. Republicans were challenged on the inhumanity issues and the feasibility. The typical response referred to the use of monkeys by scientists to the betterment of human health. In one interview, the questioner drilled down. "But in this case the being is not an animal.

The response? "We've used human beings in research before."

"The practice in the case of infecting Blacks to study cures for venereal disease is seen as a mark of shame on our history. Now experimenting on humans without their permission is no longer legal." the interviewer pointed out.

"But we are not talking about humans."

The interviewer moved on. "Your opponents argue that this is really a bill to experiment on new equipment that might not achieve the intended goal. Nothing like this has ever been done before, and there are many issues, such as how you get one of these beings into the vessel without harm or even destruction."

"That's what science is about, doing things never been done before. I am not a scientist, so I can't predict how they will accomplish each step."

A typical interviewer of those opposed to HR4922 would remind a Democratic member of the Appropriations Committee that Democrats typically advocate for scientific progress and ask, "So why is this bill any different?"

One responder answered. "There is no scientific evidence that the project is feasible. No one knows what material and thickness can withstand the pressure at the depth at which this society of human-like beings exists. The bathyscaphes on Yellow Submarine explore near the currently greatest depth, and we don't even know how much deeper they need to be able to go because Alonites live below our sea floor."

"Are you sure?"

"Yes, I've studied the reports from the ROE team. The being they've named Alon has telemessaged this information."

"Wouldn't telemessaging be of value to humans?"

"Possibly, but the idea that locating the special part of the brains of these beings is like thinking that by studying octopuses, we could grow brains on our fingertips and toes."

The interviewer laughed but continued, "Would you support funds to examine the feasibility?"

Wagging her head and making an uncertain face, she replied "I probably would, but I'm not willing to commit unless I can review the details, you know the ones where the devil lives."

When Sanjay reported the defeat of HR4922 with a poorly contained grin on his face at the next ROE meeting, cheers could have lifted Yellow Submarine out of the water. Elizabeth beamed along with him and hugged Tyrone while making eye contact with Pierre. She knew that this wasn't the demise of Jackson's plan, but she wouldn't let the fact of Jackson's persistence and resources spoil the win of the moment.

Chapter 91

(Follow the Money)

That Congress turned down HR4922 did not surprise Director Jackson. In fact, he relished the fact it gave him a stronger case for private support. Elizabeth learned he had garnered pledges amounting to one hundred million dollars to research feasibility. Jackson had signed a letter of intent to hire Ocean Deep to design a stronger submersible.

The funding constituted only a small challenge, given the actual cost of a virtual miracle to keep a captured Elonite alive during an examination. She considered a submersible medical laboratory nothing but a fancy of wealthy men and suspected Jackson knew better. It was difficult to believe that Jackson believed that a means of inspecting an Alonite brain could be accomplished without killing the victim, and she didn't believe he cared.

Elizabeth's increasing ability to communicate with Alon allowed her to reassure him that he and his society were safe for a considerable amount of time. To further guarantee his and fellow Alonites, she advised that if he could convey everything known about the ability to telemessage, Jackson's reason for capturing would become pointless. Further, with the help of Pierre, Tyrone, Heather, and Haihong, she concocted a list of questions that Jackson or any scientist might have.

* * *

Sanjay kept Command Center up-to-date with Elizabeth's considerable progress in communicating with Alon, and asked the committee to review Elizabeth's list of questions they might want to know about the Alonite society, including questions about both the hard and social sciences. The unexpected lukewarm response puzzled him, but they did agree to consider constructing a list. Jackson chided him on using the phrase 'hard science' since it connoted that the social sciences were soft science, even though Command Center contained no social scientists. The response did not surprise Elizabeth. She believed Jackson would try to throw a monkey wrench into any plan other than his own to learn more about Alonites.

Chapter 92

(Elizabeth garners information on Alonites)

Tyrone no longer piloted explorations on the caterpillar. The near disaster haunted him not because of the dance with death, but because he felt responsible for the near disaster. Further, he trusted Pierre's piloting, and now felt comfortable speaking up in meetings and working with others. Whether it was a new lease on life or it was therapeutic to relate his story about the white woman he lost to non-racist racism. He didn't know.

Of course, Elizabeth had no objection to working alone with Pierre. In fact, it was win-win as Pierre seemed particularly astute regarding communicating with Alon.

They found they enjoyed getting involved in philosophical and psychological discussions. On the current trip, Elizabeth said at the outset, "Have you ever noticed that our return to Yellow Submarine always seems shorter than our venture out ?"

Pierre frowned slightly before responding. "I think you're right. On hikes or trips in a car, the same seems true. It must be psychological. It also seems relative to what you're doing." He tipped his head toward the caterpillar's ceiling in thought.

"Alvin Toffler said that humans measure time as a ratio of how old you are," Elizabeth said. "If a thirty-six-year-old mom tells a twelve-year-old daughter she can wear lipstick at thirteen, their sense of the wait is different. One year is 1/12th or 3/36th of the time the daughter has lived, while it's 1/36th of Mom's lifetime. The daughter sees it as three times as long. I wonder if that's related."

Pierre appeared to like the challenge. "So before we leave Yellow Submarine, we have out and back ahead of us, while on the return, we only have the back ahead of us. I guess that makes sense since the way back is one-half of the trip we set out to make." Pierre tightened his lips, looked up, and nodded slightly. After a pregnant pause, he said. "I love being alone with you." He looked up at the mirror and directed his soft eyes on her face.

"And I, you," she responded. "I'm sorry I've been so distracted lately that I've practically ignored you."

"That's fortunate. I would not be able to forgive myself if our feelings for each other caused you to be banned from the Yellow Submarine project. Jackson would leap at discovering a valid reason to dismiss you."

"You are so magnifique!" Elizabeth brought her fingers to her lips and blew him a kiss just before her jaw dropped. "Oh, I see Alon's scaphe!!" she exclaimed. " I was afraid we'd never see him again."

Alon telemessaged he had taken a break from his job and described the caves he loved to explore before asking for the symbols for the list of words prepared by the Blues.

It did not take long to put up the letters for the list, which Elizabeth recognized as ones they'd need to warn Alon of an imminent attempt to capture him. For the first time, she felt completely confident that Jackson could not succeed.

Chapter 93

(The ROE team weighs in on topics relevant to Alonite life)

After Elizabeth completed her dictionary and record of everything she'd learned about Alonite society beyond their governance, she shared it at the next team meeting. Given that each member of the team specialized in a different aspect of science and technology, they would be capable of forming a list of words that Elizabeth could use to query Alonite's understanding of the different areas of science.

Sanjay called for the team to turn in their lists to the situation room. He kept a pocket nailed next to his door for people's suggestions.

After the meeting ended, Alessandra approached Elizabeth. "I understand that Pierre is piloting your caterpillar these days."

"That's right," Elizabeth said curiously.

"You know, I've always thought the caterpillar too confining, but I'm having second thoughts. I think I'll check it out."

"It would be good if more people experienced Alon's telemessaging," Elizabeth said unenthusiastically, "but …"

"But, what?"

"Oh, nothing."

Once Sanjay had received all of the lists, he passed out a compilation of the scientific words to people's bedrooms, announcing that they would meet after the evening's dinner.

Sanjay began the meeting with "I assume you've all seen the lists needed to speak the language of our individual science specialties and compared them with the list of new words Elizabeth has taught Alon related to human society. Obviously, it would be impossible to determine everything that Alonites know, but is there anything else that is worthwhile pursuing? This information will be communicated to Command Center in order to discourage Director Jackson from his ambitious goal. In fact, let me rephrase the goal. There's nothing wrong with Director Jackson finding areas where he wants more information if he believes that further communication between Elizabeth and Alon could provide it. Therefore, my question to you is whether the report provides a good overall picture."

Before Sanjay could finish, Yusuf interrupted. "The one thing Director Jackson wants above all is to learn how Alonites can telemessage. We seemed to have wandered from his goal."

"You have a valid point, but this report will go to all members of Command Center. Even if they approved Jackson's initial plan to seek a way to learn about how Alonites telemessage, they have the ultimate power to determine how humans will interact with Alonites going forward."

Yusuf countered, "Going forward? That's an overused phrase from the past. No one uses it anymore."

"Oh my God, Yusuf, who anointed you the keeper of our word choices?" Hans asked, and to his delight, the rest of the team snickered.

Elizabeth had planned to remain quiet, but pleaded, "We are getting way off course. Sanjay asked whether the report provides a good overall picture of the science humans know and whether it corresponds to Alonite's body of science, not the situation's politics.

"I would like to know if Alonites eat anything besides fish. Are there plants they find edible?" Hans asked.

"That's a good question. Elizabeth, I assume you are keeping track of relevant questions," Sanjay asked.

"I'm keeping track of all of them. I'll sort them later."

Hans's question motivated other suggestions.

"I'd like to know more about this infothon Alon carries around that the government uses to communicate. They sound like devices no one can live without, even more so than humans and their smartphones," Haihong added.

"I think it's interesting that their God is an ancient lobed fish. It sounds like a coelacanth, from which humans are believed to have evolved," said Heather. "I'd just like the report to make a note of this. I think it's an interesting connection between humans and Alonites."

"Interesting point," Sanjay said. "The more Alonites seem like humans, the better. I wonder if they also evolved from coelacanths, as some scientists believe we do. Christians believe humans are made in God's image, and, interestingly, Coelacanth is the Alonite god."

"They seem to ignore the existence of laws of gravity and are more in tune with water pressure and wave strength, which is completely understandable," Tyrone offered.

"Hardly understandable," Yusuf said, again without being recognized by Sanjay.

Otherwise, the discussion stayed on course with several other minor suggestions. Elizabeth waited for Pierre and walked out of the dining room with him. "That went well." Her eyes told him what her mouth had never said, namely that she loved him. Before they parted, she said, "Oh, Alessandra wants to join us in the caterpillar."

"Oh, really?" Pierre's face was unreadable, but what man isn't attracted to such a beautiful woman?

Chapter 94

(Jackson's significant announcement)

Elizabeth had nearly finished tweaking her catalog of Alonite society, incorporating the suggestions brought up at the meeting, when Sanjay announced that Director Jackson planned a trip in Ocean Deep's latest submersible that had been tested to handle the pressure at the deepest depths of the Indian Ocean. She gulped. Even though that didn't mean the new scaphe could stand the pressure below the ocean floor, she never dreamed that things would happen so fast.

"If the claim is true, it's a far cry from what's needed to interact with Alonites," Sanjay said reassuringly focusing his attention on Elizabeth as if he had read her mind. Her chin dropped quickly to her chest and slowly back up. "Did he say when? I have to contact Alon, but we only meet at our designated times! We've already met today." Everyone's faces wore expressions of concern, more over Elizabeth's reaction than over what it meant for Alonites.

Alvin Chapter 95

(Alvin and Haggard encounter a problem)

Alvin and Haggard's expeditions now took a considerable amount of time to learn Elizabeth's language by guess and by luck. Alon worried they'd be called out for not bringing in enough fish for the uppers. He shared what he learned about Elizabeth's species with Haggard and often invited him to help with guessing Elizabeth's word meanings. He often thought about Professor Thurgood and wished he were still alive. How he would love to see his face when he learned that his action against the government not only ultimately provided a better life for his followers in the country of the Blues, but enabled Alvin's subsequent ability to telemessage. This, in turn, resulted in communication with the beings living in the upper world. Alvin wondered if it were fate or luck.

This particular dek, Alvin and Haggard headed out for their expedition. When they punched the fan controls, instead of the fan stopping, folding up, and retracting down the tunnel opening, they heard the sound of a straining motor. "What the ashes!?" They waited a bit and tried again with the same result. "Now what?" Haggard asked.

"I have to go up there and see if I can find out what's wrong. Don't touch those controls while I'm up there!"

Haggard made a face. "What kind of a mudfish do you think I am?"

Alvin punched his arm. The pressure just under their ceiling was already dangerously low as too many daring teenagers had discovered the hard way over the generations. He donned his aquanaut suit.

He carefully rose through the tunnel, not knowing what he might find. The expensive fan was one of a kind, but they thought to manufacture replacement blades along with a supply of appropriately sized nuts and bolts. It had clearly stopped operating; otherwise, he would be moving against the clean water blasting downward. He proceeded cautiously, reminded of the work it took to straighten the tunnel so the fan could more forcefully blow down the clean aqua. It was also much easier to travel through without the winding passageway he and Fitz once had to navigate inside the aquascaphe.

Soon, he could see the bottoms of the blades. They were not moving, and then he saw the problem. The blades held an aquascaphe larger than the one that Elizabeth occupied. It was stuck between two bent blades. It must have come too close and was sucked into the fan, just like the many fish that got too close. The fans had to be regularly cleaned of the mutilated bodies.

"Ashes!" Alvin cursed. This must be the fishass that Elizabeth had told him about. Now what? They hadn't been able to turn off the fan, and if he could push the interlopers off, the fan could become unstuck, and the whirling blades would slice him into pieces. He saw the occupants waving frantically at him, but he ignored them. He needed to get the fan's motor stopped. Removing the blades from the fan was out of the question until the fan's motor was shut down.

Somehow, he or someone needed to get at the inner works and disable the wires. He'd have to go back to the department and get a mechanic. He dropped down the tunnel as quickly as he could. It seemed to take forever for him to go through the procedure to get back into the bathyscaphe. As soon as he removed his helmet, he told Haggard to speed back to their garage. He promised to explain en route.

Upon arrival, a frustrated Alvin couldn't get through to the mechanics that he needed help immediately to see to the giant fan. He had almost forgotten that besides himself, only Fitz and Haggard knew he had encountered these living upper beings. "There's something stuck in the fan, and it won't shut down. We need a mechanic right now to get at the wiring."

"What? What's stuck?" said the head worker in the building housing their bathyscaphes.

"It doesn't matter what it is. Get someone's ass on it right now!"

"Oh, don't get your fins in a twizzle. Mikon is here," he said and then yelled for him.

Mikon came quickly from the back room. "What's up?"

"I'll tell you on the way. Grab your aquasuit, a toolbox, a heating device, and let's go."

Finally, Alvin's desperation got through, and Mikon hustled, grabbed what was requested, and swam into the bathyscaphe. Haggard remained inside, and as soon as all doors were closed, he sped back to the fan site. "Don't ask questions. Just take my word for what needs to be done. The fan won't shut off, so you'll have to go in and cut or disengage the wires at the base. I don't know how you get in there, but, if necessary, melt a hole in it."

Mikon blinked. Everyone understood that Alvin was an aquanaut hero for good reason, and he would do what he was told.

Alvin finally relaxed when they arrived back at the base of the tunnel. He demonstrated the struggling noise when he tried to shut down the fan.

Mikon removed the main panel and untwisted a nest of wires from the base while Alvin and Haggard climbed into their suits. The buzzing sound was gone when Alvin tried to shut off the valve again. He ordered Mikon to stay with the aquascaphe and asked Haggard to join him up the tunnel. He carried the toolbox with him. First, they tried manually pulling one of the blades off the foreign aquascaphe.

Pulling steadily, it began to budge, and then suddenly snapped into its proper location, pushing the invaders to the edge of a single blade. Then Alvin and Haggard easily pushed the vehicle out of the fan's way. It was scratched, but the damage was minimal. He and Haggard pushed the scaphe far enough above the fan to escape the downward thrust. They returned and ordered Mikon to restore the wires in the panel.

Alvin and Haggard agreed there would be no further explorations for an indefinite amount of time. The fan needed at least two of the blades replaced to continue sucking clean water into their country, but Alvin had to think of a reason explorations would need to be suspended once the fan was working properly. He decided it made sense to honestly relate they had encountered an aquascaphe containing live beings like the dead ones found on a transportation device long ago. No one would suspect that Alvin had been communicating with these beings. Surely, the uppers will not want to go to war with an unknown society. In any event, explorations had to stop. Now, more than ever, he wished Professor Thurgood were alive.

Chapter 96

(Jackson visits Yellow Submarine)

Director Jackson and his pilot notified Sanjay via MAV that they were on their way to stop at Yellow Submarine. When Sanjay told Elizabeth, hope dared to creep inside her. She'd have one last chance to convince Jackson to give up or delay his plans.

Sanjay called an emergency meeting to discuss last-minute ideas about discouraging Jackson from searching for Alon. As soon as he conveyed the news, Yusuf spoke up. "Why discourage him? Am I the only one with the mind of a true scientist? Despite all that Elizabeth has discovered, studying the makeup of Alon's body would tell us so much more."

While Elizabeth's blood came to a simmering boil, everyone else began to talk at once.

"You're the only one who is crackers enough not to respect the rights of another species not to be subjected to experimentation," said Heather.

"Why don't you volunteer to have your body examined by the Alonites," Hans said.

"Why isn't all that Elizabeth discovered enough?" Tyrone asked.

"Yes, why do we care how many bones Alonites have or the shape of their skull? asked Pierre.

"Australians may have treated Aboriginals cruelly, but they never performed surgery on their bodies for the sake of science," Sidney said.

"Why do not you volunteer to be physically examined by Alonites in exchange for one of them?" gentle Haihong said, surprising everyone she had been so bold.

"You are proof that scientists can be idiots outside of their specialties," Alessandra said.

"We are all tired of your arrogance," Haru said. Haru usually avoided conflict.

Elizabeth barely suppressed her temptation to smirk at the bombardment by the ROE team, and Sanjay, who normally would have called for order, said nothing. Yusuf stormed out of the meeting, mumbling, "You people are unbelievable."

Sanjay said, "Well, that was interesting. Can we now calmly discuss how to approach Director Jackson?"

"I think you should take the lead," Alessandra said, "and talk about how we agree there's no need for him to seek out Alon because of how much our darling Elizabeth has uncovered."

Elizabeth winced. She had noticed Alessandra's interest in Pierre. Was she also jealous of the attention Elizabeth received at the discovery and relation with Alon? Most likely. Her beauty made her the center of attention all of her life.

Haihong and Heather also seemed startled by Alessandra's snarky remark and looked at Elizabeth to see her reaction.

"I agree Sanjay is the best spokesperson," Elizabeth said, determined to ignore any untoward comments. I'd like to be involved as little as possible, but if asked, I will outline my report."

"How are we going to manage Yusuf?" Tyrone asked.

"We don't," Pierre said, "but make it clear he stands alone."

"Let me deal with him," Sanjay said. "I want you to know that what just happened will not be tolerated again. We need to work together. I hope this was therapeutic, but it cannot be repeated. I will speak with Yusuf in a way that is both apologetic and explanatory."

"Yes, we understand," said Heather, "but please help him see that he is among intellectual equals, something to which he is unaccustomed."

Sanjay nodded. "Back to work. Jackson will be here soon."

Chapter 97

(Jackson stops at Yellow Submarine)

Sanjay returned to the situation room, barely in time to receive the signal that Director Jackson was just outside of Yellow Submarine's port, and that his remote control to open the delivery doors wasn't working. He set the outer entrance door to open and then tried to curb his curiosity. On his way down to the delivery level, he heard the scaphe entering and, soon after, the door closing. He remained waiting for the ten minutes it took to siphon off the ocean water.

A shocked Sanjay froze to see Director Jackson struggling to open his severely dented door with scratches and scrapes along the sides of the scaphe. He couldn't imagine what they could have hit that would cause that much damage. He hastened to assist, but the door was too bent. "I'll get help," he said and rushed off. He found Tyrone and asked him to round up every mechanic on board. "The scaphe is damaged, and they'll need something like a jaws-of-life." He returned to Director Jackson and told him help was on the way. If Director Jackson was concerned, he did not show it.

Sanjay tried to communicate with him through the several extra thick plates of transparent material, but the director shook his head. Which Sanjay assumed he couldn't hear or wanted to wait to explain.

Tyrone and two mechanics turned up with jaws-of-life, crowbars, and metal cutters.

Sanjay said, "Please find a way with the least damage to get this door open."

"Well, it can't be done without damage," Tyrone said, "and the director and his pilot aren't going anywhere in this vehicle for any time soon. They'll have to take the spare caterpillar back home, and Ocean Deep may have to return to fix it. Or they can tow it back, but that's their call."

Chapter 98

(ROE team anticipates Jackson's story)

News traveled through Yellow Submarine faster than a mouse skittering across a kitchen floor. Director Jackson refused to talk about it until Sanjay assembled everyone.

"Let's do that over dinner then," Sanjay said. "You may need to relax for a bit. We have spare beds."

"No, thanks, but I need a drink and a quiet place."

"Done and done. I'll clear the library and grab a bottle of wine and two glasses," Sanjay said, nodding at the pilot.

As soon as he had settled the two men, Sanjay located Elizabeth. After he conveyed the state of Jackson's scaphe, she said, "I know exactly what happened. I told you that a giant fan blocks the entrance to Alon's country. They must have run into it. It runs constantly to send clean water down into Alvin's polluted space. It's also the passageway between the bottom of our ocean and the top of theirs. Alon must retract it to pass through every time he explores."

"Oh, right. You've mentioned that fan before," Sanjay said.

"I wonder how he feels now about capturing an Alonite," she said, feeling disgust pouring from every cell. "I can't imagine there will be any Alonite aquanauts leaving their country ever again. This is one for their history books." Suddenly, her shoulders drooped. She didn't know if she was upset about the near disaster or delighted at Jackson's huge failure.

Sanjay said nothing, waiting for Elizabeth to collect herself. When she had wiped tears from her eyes and had calmed down, she said, "Sorry."

"There's nothing to be sorry about, Elizabeth. Things have certainly suddenly become interesting. I don't know what to expect for this evening's meeting. You may want to take a nap."

"Are you kidding? There's no way I could sleep."

"Let's go to the fifth floor and take a walk then."

Chapter 99

(Jackson addresses ROE team)

Hans never failed to rise to the occasion. He could make a delicious meal from the leftover scraps in a compost bin. Fortunately, there were always chickens and many ways to prepare them. He used only breasts and oven-roasted the other parts for another day. He sliced slits into the breast meat and inserted cheese and baby spinach from the nursery. He dipped them in beaten eggs, floured them, and pan-fried them. Boxed brownie mixes from Ghirardelli in San Francisco needed only eggs and water for a tasty dessert.

Unfortunately, the circumstances were such that people did not notice what they were eating. Hans understood, but his face displayed his disappointment.

After the tables were cleared, Sanjay stood. Director Jackson sat at a nearby table. "Director Donald Jackson and his pilot have experienced some accident with the new aquascaphe. I will let him describe what happened."

Jackson rose. "As most of you know, it is in the best interest of science to understand as much as possible about the newly discovered species. I have led the cause precisely to do that. Despite resistance at many levels, I succeeded, together with the company Ocean Deep, in creating a two-compartment bathyscaphe strong enough to withstand the enormous pressure at the bottom of our oceans, including the Indian Ocean, our deepest. We merely descended for proof of concept. I want to emphasize that our scaphe passed easily, and we were almost at the ocean's bottom. Our scaphe is not damaged from the pressure, but from a giant fan we encountered, no doubt placed by Alonites. It sucked us into its blades. It did dent the scaphe, but the exterior obviously held, or we wouldn't be here. We were trapped there for hours before …" He hesitated. "Before two presumed Alonites in suits much like those of our ocean explorers arrived and managed to push us free. Once we were free, we could navigate, so evidently, the working equipment was not damaged." We considered it prudent to stop here. Needless to say, we were shaken, but" he held up a glass of wine "a little relaxation has worked wonders."

After Jackson finished, Sanjay asked. "Are you willing to answer questions?"

He looked displeased but said gallantly, "Of course. What is it you want to know."

"I'll begin," Sanjay said. "What are your repair plans for the scaphe? You could return using our spare scaphe and send experts from Ocean Deep down here to do the repairs or tow it back."

The director looked to his pilot, who stood and said, "I work for Ocean Deep, but that decision should be made by upper management so Chair Jackson and I will not be towing it back."

"Other questions?" Sanjay asked.

Heather stood. "Yes, Elizabeth may be too modest to talk about what she's learned by teaching Alon to understand written words on a whiteboard. She has just completed a report that could answer all the questions any scientist, including social scientists, might ask. That means there is no need to capture an Alonite."

"Does she know what about their brain structure allows them to telemessage?" Director Jackson's shoulders straightened.

"Let me answer that," Elizabeth said.

"Their ability is complicated. Very few have that capacity, and those that do have to be taught at birth or be treated with a machine before intense practice. A secret society treated Alon, and he knows nothing about the machine used. Undoubtedly, he was one of the Alonites that saved you, and neither he nor any Alonite will ever leave their country again."

The director bristled. "Why do you say that?"

"Because I've informed him of your desire to capture him and study his brain. And now he knows you have vehicles that survive at the pressure just above their country."

"What!! How could you?"

Sanjay looked aghast and gave Elizabeth a 'your goose is cooked' look.

Pierre stood. "Your quest is next to impossible now anyway. Alonite aquanauts won't be exploring above their ceiling after encountering you. Elizabeth's report states that they've explored as high as they can go, and they're just bringing back uncontaminated fish for the upper class. Their society is much like ours. Corporations have polluted their environment and food but won't act except to keep their individual domains as pure as possible."

Haru blurted out, "I want to address the pilot from Ocean Deep." He looked directly at the pilot. "Do you know how much pressure the scaphe can withstand?"

"We know it can easily function at the very bottom of the ocean floor."

"That's not good enough," Tyrone spoke up. "The Alonites live much deeper. In fact, they suffer in pressure that they deem too low at the bottom of our ocean, and even just below our floor, their ceiling. You noted they were wearing suits."

Director Jackson looked as if a lightning bolt had hit him. "How much deeper is their floor?"

"It's in Elizabeth's report," Sidney said in a smug singsong tone.

Haihong had had her hand raised while everyone else had broken in. Sanjay recognized her. "I only want to say that I agree with my colleagues and urge you to read Elizabeth's amazing report."

"I'm very tired," Director Jackson said. I need to sleep on all of this. Please excuse me."

Yusuf raised his hand, causing Sanjay to look at the director quizzically. When given a reluctant expression of approval, he nodded at Yusuf.

"I want us all to give Director Jackson a big hand for all his efforts to learn more about this species even if they did not turn out." He began to clap vigorously while the rest of the team pat-a-caked with fingers to palm. Sanjay and Pierre struggled to keep the smiles from their faces, presumably because Yusuf was unaware that he may have gently closed the coffin on Director Jackson's Alon project.

Epilogue

Director Jackson gave up on pursuing an Alonite and did not act on the impulse that burned inside him to suspend Elizabeth a second time. Instead, he notified Command Center that he would resign in a few weeks. He had purchased Ocean Deep, and he planned to turn the company into one for tourists. Ocean Deep would advertise one tour to the Titanic and another to the entrance to Alonite land blocked by a giant fan. His ads would hint at the possibility of seeing an Alonite. He would contribute more to his family's fortunes than any other Jackson.

Alvin accepted that his authoritarian government of Pelagia would never change and, along with Haggard, joined the Blues in Abyssal. Alvin regretted he couldn't risk inviting Fitz. Alvin settled in the section of the capital city inhabited by former Pelagians. He became a teacher of the history of Pelagia and is seeking a mate.

Command Center praised the thoroughness of Elizabeth's fascinating report. When the report's existence was leaked to the public, major book publishing houses vied to turn it into a book, certain it would be a best seller. Command Center gave Elizabeth time off to collaborate with an editor in exchange for 35% of the royalties. She accepted. She and an excellent content editor finished working full-time in five weeks.

Elizabeth never saw or heard from Alon again, nor did she expect to. She and Pierre often returned to the meeting coordinates at the agreed-upon time with low expectations. In fact, like Alon's explorations, their trips devolved into fishing expeditions ever since they adopted the Alonite practice of attaching a net to the caterpillar.

The team settled into its usual routine and anticipated declaring that phase one of the experiment had been successful overall. Even farmers experienced issues with their crops, and the nursery problems on Yellow Submarine were deemed manageable. They compiled a list of desired improvements and convinced Command Center to abandon Rule # 9 — the no sex and no marriages rule — in testing whether submarine life could be tolerated in the long term. In addition, they recommended the beginning of phase two with a larger submarine comprising families and children.

When Elizabeth's book came out only her friends understood the irony in Elizabeth's dedication. "This book is dedicated to the ROE team with which I worked on Yellow Submarine and Director Donald Jackson for inspiring me to do the necessary work to communicate with Alon. Without this inspiration, I never would have discovered so much about the Alonites."

Relevant websites

https://www.npr.org/2023/01/24/1151062783/scientists-discover-fantastical-creatures-deep-in-the-indian-ocean

https://oceanservice.noaa.gov/welcome.html

https://www.britannica.com/science/standard-atmosphere-unit-of-measurement

About the Author

Eloise Hamann is a retired Professor of Mathematics from San Jose State University. Her passion for puzzles led her to become a mathematician, and her passion for imagining and justice led her to become a writer and political activist after retiring.

She writes realistic, character-driven speculative fiction with a message about caring for our environment.

This is her fourth book. She has written several short stories and poems published in collections of work by many authors. They can be found on her web page https://ewrites.wordpress.com and by searching for her name on Amazon Books.

She also self-published her son's novel after his death and entered a collection of his poetry, *Deathdoubledactyl* in a contest that resulted in publication.

Other Books by the Author & Deceased Son

This is not your ordinary thriller. It's Stephen King meets Garrison Keillor. A small town cop's peaceful life is blown apart by a series of murders in his beloved Dry Creek. He must not only determine whodunit, but *how* they did it. The novel track's Ben's struggle to accept the awful truth.

Landley, a lower caste member of a society dwelling under the ocean floor, is accidentally associated with a clandestine association, whose mission is to ensure tetraped survival. The powerful empire refuses to deal with a critical hazard to its citizens. The paranoid leader sees criticism as treason. When hints of the group's existence surface, the members' lives are threatened. Landley's heartthrob detective, Lawson, desperately fights to protect them.

Set in the deep sea, Landley & like-minded tetrapeds fled from a government planning to kill them. The human-like human beings struggle to survive in the primitive Lost Sea.

A hysterical tale by Shannon Hamann, deceased son of Eloise Hamann, about an insanely audacious, pathologically dishonest young woman, who hangs out her shingle as a therapist, without training or qualifications. The title refers to an obsession she has with Brad Pitt.